T0131048

Comparing the Organizational Cultures of the Department of Defense and Silicon Valley

NATHAN VOSS, JAMES RYSEFF

Prepared for the Office of Net Assessment
Approved for public release; distribution unlimited

NATIONAL DEFENSE RESEARCH INSTITUTE

For more information on this publication, visit **www.rand.org/t/RRA1498-2**

About RAND

The RAND Corporation is a research organization that develops solutions to public policy challenges to help make communities throughout the world safer and more secure, healthier and more prosperous. RAND is nonprofit, nonpartisan, and committed to the public interest. To learn more about RAND, visit www.rand.org.

Research Integrity

Our mission to help improve policy and decisionmaking through research and analysis is enabled through our core values of quality and objectivity and our unwavering commitment to the highest level of integrity and ethical behavior. To help ensure our research and analysis are rigorous, objective, and nonpartisan, we subject our research publications to a robust and exacting quality-assurance process; avoid both the appearance and reality of financial and other conflicts of interest through staff training, project screening, and a policy of mandatory disclosure; and pursue transparency in our research engagements through our commitment to the open publication of our research findings and recommendations, disclosure of the source of funding of published research, and policies to ensure intellectual independence. For more information, visit www.rand.org/about/principles.

Published by the RAND Corporation, Santa Monica, Calif.
© 2022 RAND Corporation
RAND® is a registered trademark.

Library of Congress Cataloging-in-Publication Data is available for this publication.
ISBN: 978-1-9774-0965-2

Cover designer: Carol Ponce; Photos: Jacob Lund/Adobe Stock, ktsdesign/AdobeStock, BillionPhotos.com/ AdobeStock.

About This Report

The ability to leverage artificial intelligence (AI) technologies and capabilities is viewed as vital to the long-term success of the U.S. military. Although there is a common understanding of the importance of AI, differences in organizational cultures and mindsets may leave the U.S. Department of Defense (DoD) unable to fully leverage the full range of capabilities developed by the private sector. Consequently, this study maps the organizational cultures of both DoD and Silicon Valley software companies to determine where they have substantial differences and where the two communities might find common ground.

This research was sponsored by the U.S. Department of Defense's Office of Net Assessment and conducted within the Acquisition and Technology Policy Center and the Forces and Resources Policy Center of the RAND National Security Research Division (NSRD), which operates the National Defense Research Institute (NDRI), a federally funded research and development center sponsored by the Office of the Secretary of Defense, the Joint Staff, the Unified Combatant Commands, the Navy, the Marine Corps, the defense agencies, and the defense intelligence enterprise.

The research reported here was completed in August 2021 and underwent security review with the sponsor and the Defense Office of Prepublication and Security Review before public release.

For more information on RAND's Acquisition and Technology Policy Center, see www.rand.org/nsrd/atp or contact the director (contact information is provided on the webpage). For more information on RAND's Forces and Resources Policy Center, see www.rand.org/nsrd/frp or contact the director (contact information provided on the webpage).

Acknowledgments

We would like to thank Anita Szafran for helping us compile documents for the analysis, and William Marcellino and Michael Ryan for their assistance with RAND-Lex. We also thank our reviewers Maria Lytell and Marek Posard for their helpful comments, and Kathryn Bouskill, Diana Gehlhaus Carew, Jason Etchegaray, Peter Eusebio, Stephen Gerras, Michael Linick, AnnaMarie O'Neill, Ricardo Sanchez, Naomi Shapiro, Matt Strawn, Taylor Willits, Leonard Wong, and Sarah Rebecca Zimmerman for providing helpful insights about this project.

Summary

Artificial intelligence (AI) has become widely recognized as a technology that is essential to the future of national security.[1] However, unlike previous eras, the U.S. Department of Defense (DoD) is no longer the primary driver of research and development investment in these types of advanced technologies.[2] Instead, large software companies that derive the bulk of their revenues from nondefense sources employ the greatest reservoirs of AI talent and invest the majority of capital into improving their AI algorithms. Consequently, DoD has sought to collaborate more effectively with the software companies of Silicon Valley.[3]

However, differences in organizational cultures and mindsets may leave DoD unable to leverage the full range of capabilities developed by the private sector.[4] To understand these differences, the authors of this report explored the values and traits that each community has sought to instill into its members. We did this by investigating how influential individuals in each community talk about the culture that they try to instill in their organizations and by categorizing documents that establish organizational culture, such as Netflix's culture deck, Amazon's leadership principles, and manuals and creeds from the military services.[5] We mapped how the language used in these documents correlated with concepts and ideas from five prominent, relevant, and highly researched organizational culture types: Hierarchy, Adhocracy, Market, Clan, and Sense of Duty.

We found that the organizational cultures that these communities attempt to instill in their members have areas of wide divergence and areas of greater commonality. In particular, technology companies put minimal emphasis on elements of Hierarchy culture and Sense of Duty culture, while military culture emphasizes these traits to a much greater degree. More promisingly, both communities embrace elements of Clan culture, Market culture, and

---

[1] National Security Commission on Artificial Intelligence, *Final Report of the National Security Commission on Artificial Intelligence*, Arlington, Va., March 2021.

[2] J. Sargent and M. Gallo, *The Global Research and Development Landscape and Implications for the Department of Defense*, Washington, D.C.: Congressional Research Service, R45403, June 28, 2021.

[3] We use the term *Silicon Valley* to refer not to a geographic place but rather to a culture and a mindset typically used for U.S. software firms employing highly educated, innovative, and skilled engineers. This includes such companies as Microsoft and Amazon (both headquartered in Seattle), San Francisco-based corporations (such as Alphabet or Facebook), and start-ups and midsize companies in a variety of U.S. cities (see C. Metz, "Pentagon Wants Silicon Valley's Help on AI," *New York Times*, March 15, 2018).

[4] National Security Commission on Artificial Intelligence, *Final Report of the National Security Commission on Artificial Intelligence*, Arlington, Va., March 2021.

[5] DoD is a large and complex organization with numerous suborganizations. For this study, we focused our analysis on the organizational culture of four military services—the U.S. Army, Navy, Air Force, and Marine Corps. Suborganizations within DoD, such as the Defense Advanced Research Projects Agency or the National Security Agency, will have their own distinct organizational cultures, which vary from the overall culture of DoD to a greater or lesser degree.

Adhocracy culture. Recognizing and building on these areas of commonality would better enable DoD to cooperate with leading technology and software companies and improve DoD's ability to recruit and retain AI talent.

Contents

Figures and Tables

Figures

Tables

CHAPTER ONE

Introduction

Making military advances in artificial intelligence (AI) has become an important goal for many of the world's leading countries.[1] For the United States, in particular, the ability to effectively leverage AI technologies and capabilities is increasingly viewed as vital to the long-term success of the military and the nation.[2] As the National Security Commission on Artificial Intelligence explains, "we fear AI tools will be weapons of first resort in future conflicts," and it predicts that these weapons are likely to proliferate to all types of future adversaries because of AI's dual-use and open-source nature.[3] Despite the importance of making advances in AI, there are concerns that the U.S. Department of Defense (DoD) is not well positioned to optimally engage with the large U.S. technology companies of Silicon Valley[4] and to recruit top AI talent from U.S. technology companies.[5] Although differences in organizational culture are often cited as one of the reasons for this AI talent gap,[6] no studies have yet empirically compared the organizational culture of the U.S. military with that of Silicon Valley and fully discussed how cultural differences might affect DoD's ability to interface with U.S. technology companies and recruit AI talent. Accordingly, the goal of this project is to compare the

---

[1]   R. Waltzman, L. Ablon, C. Curriden, G. S. Hartnett, M. A. Holliday, L. Ma, B. Nichiporuk, A. Scobell, and D. C. Tarraf, *Maintaining the Competitive Advantage in Artificial Intelligence and Machine Learning*, Santa Monica, Calif.: RAND Corporation, RR-A200-1, 2020.

[2]   M. Kepe, "Considering Military Culture and Values When Adopting AI," *Small Wars Journal*, June 15, 2020; and Select Committee on Artificial Intelligence of the National Science and Technology Council, *The National Artificial Intelligence Research and Development Strategic Plan: 2019 Update*, Washington, D.C.: Executive Office of the President, 2019.

[3]   National Security Commission on Artificial Intelligence, *Final Report of the National Security Commission on Artificial Intelligence*, Arlington, Va., March 2021.

[4]   We use *Silicon Valley* to refer not to a geographic place but rather to culture and mindset typically used to refer to U.S. software companies employing highly educated, innovative, and skilled engineers. This includes such companies as Microsoft and Amazon (both headquartered in Seattle), San Francisco-based corporations (such as Alphabet or Facebook), and start-ups and midsize companies in a variety of U.S. cities.

[5]   J. Ryseff, "How to (Actually) Recruit Talent for the AI Challenge," *War on the Rocks*, February 5, 2020; and E. B. Kania and E. Moore, "Great Power Rivalry Is Also a War for Talent," *Defense One*, May 19, 2019.

[6]   A. Mehta, "Cultural Divide: Can the Pentagon Crack Silicon Valley?" *Defense News*, January 28, 2019.

organizational culture (i.e., the shared values, beliefs, and assumptions of the members of an organization or service) of the U.S. military with that of U.S. technology companies.

To accomplish this, we compared the organizational cultures of DoD and Silicon Valley using (1) a review of the extant literature and other sources and (2) a text-analytics approach to examine a large corpus of documents that contain insights about culture and values. We conclude by discussing the relevance of this work to DoD and how further studies could build on this work. Although we report a text analytic study comparing DoD and Silicon Valley organizational culture, we view this project as a preliminary step for better understanding this topic and one possible lens through which to examine this issue.

Organizational Culture

Culture serves as a powerful, intangible force in both civil and military organizational settings.[7] Although numerous definitions of *organizational culture* exist, the term broadly refers to the "set of beliefs, values, and assumptions that are shared by members of an organization."[8] A more comprehensive definition of *organizational culture* that has been proposed defines it as "a pattern of shared basic assumptions learned by a group as it solved its problems of external adaptation and internal integration, which has worked well enough to be considered valid and, therefore, to be taught to new members as the correct way to perceive, think, and feel in relation to those problems."[9] Organizational culture typically permeates throughout an entire organization, implicitly influences members' behaviors, and often serves as a critical contributor to organizational effectiveness.[10] According to Schein, organizational culture consists of multiple conceptual levels of analysis that vary in their accessibility but are each

---

[7]   P. R. Mansoor and W. Murray, eds., *The Culture of Military Organizations*, New York: Cambridge University Press, 2019; and E. H. Schein, *Organizational Culture and Leadership*, San Francisco, Calif.: Jossey-Bass, 2010.

[8]   L. S. Meredith, C. S. Sims, B. S. Batorsky, A. T. Okunogbe, B. L. Bannon, and C. A. Myatt, *Identifying Promising Approaches to U.S. Army Institutional Change: A Review of the Literature on Organizational Culture and Climate*, Santa Monica, Calif.: RAND Corporation, RR-1588-A, 2017; and B. T. Gregory, S. G. Harris, A. A. Armenakis, and C. L. Shook, "Organizational Culture and Effectiveness: A Study of Values, Attitudes, and Organizational Outcomes," *Journal of Business Research*, Vol. 62, No. 7, July 2009.

[9]   Schein, 2010.

[10]   J. A. Chatman and C. A. O'Reilly, "Paradigm Lost: Reinvigorating the Study of Organizational Culture," *Research in Organizational Behavior*, Vol. 36, 2016; C. A. Hartnell, A. Y. Ou, and A. Kinicki, "Organizational Culture and Organizational Effectiveness: A Meta-Analytic Investigation of the Competing Values Framework's Theoretical Suppositions," *Journal of Applied Psychology*, Vol. 96, No. 4, July 2011; C. A. Hartnell, A. Y. Ou, A. Kinicki, D. Choi, and E. P. Karam, "A Meta-Analytic Test of Organizational Culture's Association with Elements of an Organization's System and its Relative Predictive Validity on Organizational Outcomes," *Journal of Applied Psychology*, Vol. 104, No. 6, June 2019; T. Schmiedel, O. Müller, and J. vom Brocke, "Topic Modeling as a Strategy of Inquiry in Organizational Research: A Tutorial with an Application Example on Organizational Culture," *Organizational Research Methods*, Vol. 22, 2019.

still components of the larger organizational culture concept.[11] At the most abstract level are the deeply embedded assumptions that explain why organizations do things as they do. At the next level are the stated values, beliefs, and attitudes of an organization that are reflections of its assumptions. At the most concrete level are artifacts, which are the observable realizations of organizational culture and assumptions (e.g., company policies, jargon, rules, products, office layouts, rituals, dress codes). Furthermore, organizations might differ in the strength of their cultures.[12] For example, an organization with a strong culture might have high agreement among its members regarding cultural norms, high alignment between its culture and actual organizational practices, or values that are deeply held among its members. Although organizations often possess a dominant culture, it is possible for various subcultures to exist simultaneously.[13]

Definitions and Project Scope

This project examines the dominant organizational cultures of DoD and Silicon Valley. Although DoD is a large and complex organization—with many employee types (e.g., civilian, contractor, and uniformed military personnel), relevant agencies and subunits (e.g., Defense Advanced Research Projects Agency [DARPA], National Security Agency [NSA], Defense Digital Service)—we focused our analysis on examining DoD military branches of service for this study (e.g., Army, Navy, Air Force, Marine Corps). Because of this more narrow focus, other specific subcultures that may be present within the DoD are not accounted for within the study.[14]

---

[11] Schein, 2010.

[12] Chatman and O'Reilly, 2016. There is often a distinction between organizational culture and organizational climate. Although there is conceptual overlap between these terms, culture typically emphasizes more-holistic values and norms, whereas climate emphasizes the shared perceptions of a specific aspect of the work environment (e.g., safety climate). Additionally, culture tends to prescribe what behaviors are acceptable (i.e., rewarded or punished), whereas climate tends to be more descriptive and does not have normative implications (for a more detailed discussion, see Chatman and O'Reilly, 2016); C. Ostroff, A. J. Kinicki, and R. S. Muhammad, "Organizational Culture and Climate," in I. B. Weiner, ed., *Handbook of Psychology*, 2nd ed., Hoboken, N.J.: John Wiley & Sons, 2013.

[13] Hartnell et al., 2019; J. Martin, *Organizational Culture: Mapping the Terrain*, Thousand Oaks, Calif.: SAGE Publications, 2002.

[14] We acknowledge that discussing the "culture of DoD" or the "culture of Silicon Valley" requires making certain generalizations that do not fully capture the cultural nuances (e.g., subcultures) that exist within each of these organizations/industries, even with the more narrowed focus on the branches of services employed in this study. To communicate about culture in a useful way, however, generalized language is needed—and is in many ways necessary. Moreover, given how few empirical studies there are comparing DoD and Silicon Valley culture, we believed that it was best to first assess culture more broadly before examining specific DoD and Silicon Valley subcultures (e.g., DARPA). We fully recognize, however, that subcultures are important to examine and an excellent way for future research to expand on this initial

For the Silicon Valley corporations, we focused on Silicon Valley's more recent usage in referring to innovative technology companies whose competitive advantage derives from creating computer hardware and software rather than the term's original definition as referring to technology companies from a specific geographic location near San Francisco. In particular, we focused on five companies to analyze: Amazon, Facebook, Google, Microsoft, and Netflix.[15] These companies are some of the largest software companies in the United States, and they have made some of the greatest investments in cloud computing and AI, two technologies of particular interest for DoD. Consequently, collectively analyzing their organizational culture will shed light on the organizational culture that shapes the workplace for many software engineers and AI experts working in the United States.

Theory of Organizational Culture

Although many theories (e.g., Denison Organizational Culture Survey, Organizational Culture Profile) and measures of organizational culture exist, we relied primarily on the Competing Values Framework (CVF) as an organizing framework for this project given its wide usage, empirical support, and conceptual overlap with other theories about organizational culture.[16] The CVF uses a focus dimension (internal [e.g., collaboration/unity focus] versus external orientation [e.g., differentiation/competition]) and structure dimension (flexibility versus stability) to yield four classifications of organizational culture (Figure 1.1).

A flexible, internally oriented culture is a *Clan* culture. Clan cultures assume that organizational members behave properly when they trust and are committed to the organization. Such cultures value collaboration, support, and affiliation and associated behaviors, such as

---

study. Indeed, elaborating on the results presented in this report is crucial before any concrete policy recommendations can be definitively made.

[15] In October 2021, the parent company of Facebook was named Meta. The social media app is still referred to as Facebook.

[16] D. R. Denison and A. K. Mishra, "Toward a Theory of Organizational Culture and Effectiveness," *Organization Science*, Vol. 6, No. 2, April 1995; C. A. O'Reilly III, J. Chatman, and D. F. Caldwell, "People and Organizational Culture: A Profile Comparison Approach to Assessing Person-Organization Fit," *Academy of Management Journal*, Vol. 34, No. 3, 1991; T. Jung, T. Scott, H. T. Davies, P. Bower, D. Whalley, R. McNally, and R. Mannion, "Instruments for Exploring Organizational Culture: A Review of the Literature," *Public Administration Review*, Vol. 69, No.6, November/December 2009; K. S. Cameron and R. E. Quinn, *Diagnosing and Changing Organizational Culture: Based on the Competing Values Framework*, San Francisco, Calif.: Jossey-Bass, 2006; and Hartnell et al., 2019. Because the CVF represents a typology, it suffers from certain inherent limitations. For example, although the CVF can provide useful descriptions and classifications of culture, cultural strength and consensus are not well-assessed within this framework. Although other frameworks could have been used, they also would entail their own set of limitations (see Chatman and O'Reilly, 2016, for a discussion of such limitations). Thus, regardless of the organizational culture framework that is employed, this will invariably require making certain conceptual trade-offs (e.g., explanatory breadth versus depth). Overall, we felt that the trade-offs required by the CVF were consistent with the overall objective of this project. Such limitations are, nonetheless, important to acknowledge.

FIGURE 1.1
Summary of the Competing Values Framework

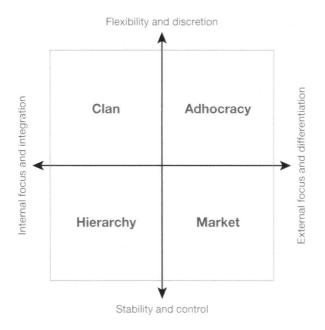

teamwork and employee involvement.[17] Organizations characterized by these cultures are often seen as effective when employees are satisfied and committed to the company.

A flexible, externally oriented culture is an *Adhocracy* culture. Adhocracy cultures are similar to Clan cultures in that they are characterized as flexible but differ in that they are more externally focused. Adhocracy cultures assume that organizational members behave properly when they view their work as meaningful and impactful. Such cultures value autonomy, growth, and stimulation and associated behaviors, such as risk-taking and creativity. Organizations characterized by these cultures are often seen as effective when employees are innovating.

A stable, externally oriented culture is a *Market* culture. Market culture is similar to Adhocracy culture in that it is externally focused but is different in that stability or control is much more valued than flexibility and discretion. They are also in many ways the opposite of Clan culture, which emphasizes flexibility and has an internal focus (see Figure 1.1). Market culture assumes that organizational members behave properly when they have clear goals and are rewarded for their performance. This culture values rivalry, achievement, and

---

[17] These culture types can emerge from bottom-up (i.e., employee-driven) and/or top-down (i.e., leadership-driven) influences. This report explores only top-down influences—the values that each organization's leadership attempts to infuse into it. These values may differ from the values that the employees collectively wish to infuse into the organization or from the lived experiences of the employees who work at these organizations. Follow-on studies to explore these two aspects of organizational culture would be a valuable supplement to this work.

competence and associated behaviors, such as being aggressive and competing with other companies. Organizations characterized by these cultures are often seen as effective when they increase their profits and market share.

A stable, internally oriented culture is a *Hierarchy* culture. Hierarchy cultures are different from Market and Adhocracy cultures in that such cultures have an external, as opposed to internal, focus. The Hierarchy culture is similar to Market cultures in that both emphasize stability, though. Additionally, Hierarchy cultures are similar to Clan cultures in that they have an internal focus. Hierarchy cultures, however, tend to value stability, whereas Clan cultures value flexibility. Hierarchy cultures assume that organizational members behave properly when there are clear roles, rules, and regulations. Such cultures value formalization, routinization, and consistency and associated behaviors, such as conformity and predictability. Organizations characterized by these cultures are often seen as effective when they function efficiently and smoothly.

To summarize, Clan cultures "do things together," Adhocracy cultures "do things first," Market cultures "do things fast," and Hierarchy cultures "do things right."[18] Although the term *competing* (as used in CVF) seems to imply that these dimensions exist in conflict with each other, research suggests that they are actually complementary.[19] Thus, both military services and technology companies can be characterized by all dimensions to some extent. However, the prominence of each dimension is expected to differ across DoD and Silicon Valley.

Organizational Culture and Recruitment

Culture plays particularly crucial roles within an organization's recruitment and retention processes. Although not usually considered, organizational culture implicitly influences these processes in many ways.[20] According to the attraction-selection-attrition (ASA) model,[21] job applicants are more attracted to organizations that value the same things that they value, organizations select employees with values that more closely align with their values, and employees with poor fit (e.g., misaligned values) will generally leave over time, which results in organizational homogeneity (e.g., employees with similar values). Indeed, a large body of research supports the notion that fit plays an essential role in the staffing process within

---

[18] K. Cameron, *An Introduction to the Competing Values Framework*, white paper, Holland, Mich.: Haworth, 2009.

[19] Hartnell, Ou, and Kinicki, 2011.

[20] M. R. Barrick and L. Parks-Leduc, "Selection for Fit," *Annual Review of Organizational Psychology and Organizational Behavior*, Vol. 6, 2019; and R. E. Ployhart, D. Hale, Jr., and M. C. Campion, "Staffing Within the Social Context," in B. Schneider and K. M. Barbara, eds., *The Oxford Handbook of Organizational Climate and Culture: Antecedents, Consequences, and Practice*, Oxford, UK: Oxford University Press, 2014.

[21] B. Schneider, "The People Make the Place," *Personnel Psychology*, Vol. 40, 1987.

both military and civil settings,[22] that high fit is linked to many positive individual-level and organizational-level outcomes,[23] and that applicants frequently use information about the organization's culture to inform their fit perceptions.[24] There is also evidence that applicants with different personality characteristics prefer certain cultures over others. For example, applicants who are high in openness to experience (a Big-Five personality trait) tend to have positive views of Adhocracy cultures but negative views of hierarchical cultures.[25] Thus, an organization's culture, or even stereotypes of that culture, will often influence which applicants are attracted to the organization.[26]

In the context of this project, if top AI talent in the private sector does not have positive views of the culture of DoD, such individuals will be less likely to accept jobs offered by DoD. Likewise, if DoD does not understand the organizational culture of Silicon Valley and in which areas that organizational culture is similar to or different from its own, it will not be in a position to optimally interface with these engineers. Accordingly, knowledge of how the cultures of DoD and U.S. technology companies are similar and different is essential for helping the U.S. military attract and retain the top AI talent with industry experience.

---

[22] Barrick and Parks-Leduc, 2019; J. R. Edwards and D. M. Cable, "The Value of Value Congruence," *Journal of Applied Psychology*, Vol. 94, No. 3, June 2009.

[23] A. L. Kristof-Brown, R. D. Zimmerman, and E. C. Johnson, "Consequences of Individuals' Fit at Work: A Meta-Analysis of Person-Job, Person-Organization, Person-Group, and Person-Supervisor Fit," *Personnel Psychology*, Vol. 58, No. 2, 2005.

[24] T. A. Judge and D. M. Cable, "Applicant Personality, Organizational Culture, and Organization Attraction," *Personnel Psychology*, Vol. 50, No. 2, December 2006.

[25] The *Big-Five* refers to the five following personality traits: Openness, conscientiousness, extraversion, agreeableness, and neuroticism. The Big-Five is one of the most widely accepted and empirically researched personality frameworks. W. L. Gardner, B. J. Reithel, C. C. Cogliser, F. O. Walumbwa, and R. T. Foley, "Matching Personality and Organizational Culture: Effects of Recruitment Strategy and the Five-Factor Model on Subjective Person–Organization Fit," *Management Communication Quarterly*, Vol. 26, No. 4, July 2012.

[26] M. E. De Goede, A. E. Van Vianen, and U. C. Klehe, "Attracting Applicants on the Web: PO Fit, Industry Culture Stereotypes, and Website Design," *International Journal of Selection and Assessment*, Vol. 19, No. 1, February 2011.

CHAPTER TWO

Comparing Department of Defense and Silicon Valley Cultures

Research suggests that there are differences in organizational cultures between industries and that industry serves as a determinant of the particular cultures that develop within an organization.[1] Because organizations within a single industry tend to have similar goals, missions, and clients, differences in organizational culture are typically greater between companies in different industries than between companies within the same industry.[2] Given that DoD and Silicon Valley are quite different along these dimensions (see below), one might expect meaningful, substantive differences in the cultures of these two organizations/industries. To explore these differences, we reviewed academic journal articles and other scholarly works referencing the culture of either DoD or Silicon Valley and consulted with subject-matter experts who had familiarity with the organizational culture of the military. This chapter incorporates the observations that these sources have made about the cultures of these two communities and synthesizes this information into initial hypotheses describing how these two cultures might compare when measured in a quantitative way along these five dimensions of organizational culture (Hierarchy, Adhocracy, Market, Clan, and Sense of Duty).

To begin with, several of the studied organizations publish *culture decks* or a statement of values that helps explain to prospective employees what kinds of people will succeed in that organization. Although it is possible that the reality of working in that organization differs in practice from these statements—as Netflix's document notes, "the real values of a firm are shown by who gets rewarded or let go"—these documents explain the type of culture that the leadership of these organization intends to create.[3] Similarly, the leadership of each of the military services publishes a statement of core values describing the expectations for service members under their command. Table 2.1 collects and contrasts these values for each of the organizations examined in this report.

---

[1]   B. Groysberg, J. Lee, J. Price, and J. Cheng, "The Leader's Guide to Corporate Culture," *Harvard Business Review*, Vol. 96, January–February 2018; B. Gupta, "A Comparative Study of Organizational Strategy and Culture Across Industry," *Benchmarking: An International Journal*, Vol. 18, No. 4, July 2011.

[2]   G. G. Gordon, "Industry Determinants of Organizational Culture," *Academy of Management Review*, Vol. 16, No. 2, April 1991.

[3]   Netflix, "Netflix Culture," webpage, undated.

TABLE 2.1

Stated Values of Military and Technology Organizations

Stated Values from U.S. Military			
DoD	**Air Force**	**Army**	**Navy**
Duty	Integrity first	Duty	Honor
Integrity	Service for self	Integrity	Courage
Ethics	Excellence in all we do	Honor	Commitment
Honor		Personal Courage	
Courage		Loyalty	
Loyalty		Respect	
		Selfless service	

Stated Values from Technology Companies				
Amazon	**Facebook**	**Google**	**Microsoft**	**Netflix**
Customer obsession	Focus on impact	Focus on the user and all else will follow	Innovation	Judgment
Ownership	Move fast	It's best to do one thing really, really well	Trustworthy computing	Communication
Invent and simplify	Be bold	Fast is better than slow	Diversity and inclusion	Impact
[Leaders] are right, a lot	Be open	Democracy on the web works	Corporate social responsibility	Curiosity
Learn and be curious	Build social value	You don't need to be at your desk to need an answer	Artificial intelligence	Innovation
Hire and develop the best		You can make money without doing evil	Respect	Courage
Insist on the highest standard		There's always more information out there.	Integrity	Passion
Think big		The need for information crosses all borders	Accountability	Honesty
Bias for action		You can be serious without a suit		Selflessness
Frugality		Great just isn't good enough		

Table 2.1—Continued

Stated Values from Technology Companies				
Amazon	Facebook	Google	Microsoft	Netflix
Earn trust				
Dive deep				
Have backbone; disagree and commit				
Deliver results				

NOTES: The Marine Corps falls within the Department of the Navy, and thus its stated values are the same as the Navy's. Similarly, Space Force falls within the Department of the Air Force. Army values retrieved from U.S. Army, "The Army Values," webpage, undated-a. Air Force values retrieved from U.S. Air Force, "Vision and Creed," webpage, undated. Navy values retrieved from Department of the Navy, "Department of the Navy Core Values Charter," webpage, undated. DoD values retrieved from Military Leadership Diversity Commission, "Department of Defense Core Values," Washington, D.C., Issue Paper No. 6, December 2009. Amazon values retrieved from Amazon Jobs, "Leadership Principles," webpage, undated. Facebook values retrieved from Meta Careers, "Culture at Meta," webpage, undated. Google values retrieved from Google, "Ten Things We Know To Be True," webpage, undated. Microsoft values retrieved from Microsoft, "Company Values," webpage, undated-a; and Microsoft, "Corporate Values," webpage, undated-b. Netflix values retrieved from Netflix, undated.

Summary of Department of Defense and Silicon Valley Cultures

Much has been written about how the culture of the military, which has a long history and deeply rooted traditions, is unique compared with a typical organization.[4] One of the defining features of the military that distinguishes it from U.S. technology companies (or any other organizations) is its overarching mission to prepare for and engage in war.[5] The military has also been described as a "greedy institution,"[6] which refers to organizations that "make total claims on their members" and "seek exclusive and undivided loyalty."[7] Such institutions make substantial physical, psychological, and cognitive demands of their members that engender especially high levels of commitment compared with a typical organization.[8]

Important values of military culture include hierarchy, loyalty, teamwork, subordination, self-sacrifice, order, discipline, mission-focus, and honor.[9] Table 2.1 lists some of the stated

---

[4] Mansoor and Murray, 2019.

[5] R. M. Swain and A. C. Pierce, *The Armed Forces Officer*, Washington, D.C.: National Defense University Press, 2017.

[6] M. W. Segal, "The Military and the Family as Greedy Institutions," *Armed Forces & Society*, Vol. 13, No. 1, 1986.

[7] L. A. Coser, *Greedy Institutions: Patterns of Undivided Commitment*, New York: Free Press, 1974.

[8] A. B. Cox, "Mechanisms of Organizational Commitment: Adding Frames to Greedy Institution Theory," *Sociological Forum*, Vol. 31, No. 3, May 2016.

[9] T. Greene, J. Buckman, C. Dandeker, and N. Greenberg, "The Impact of Culture Clash on Deployed Troops," *Military Medicine*, Vol. 175, No. 12, December 2010; and S. A. Redmond, S. L. Wilcox, S. Camp-

values of the military services. Within DoD, autonomy is not always highly valued given that military personnel are expected to obey lawful orders.[10] However, this is not to say that autonomy is not important. For instance, the battlefield can be very complex. Although soldiers are still expected to follow orders from officers with higher rank, there can be flexibility and discretion in how orders are carried out. Likewise, soldiers may need to quickly adapt to changing circumstances during combat, which could entail deviating from their roles within the chain of command and rigid organizational hierarchy. These scenarios illustrate that culture is very nuanced and complex and that multiple organizational cultures can exist concurrently. These considerations also highlight how public views of the military (i.e., as being highly hierarchical) may contrast with true military culture under specific circumstances (e.g., such as when autonomy is needed in the battlefield).[11] Thus, DoD can possess aspects of an Adhocracy culture but still be characterized as hierarchical. This characteristic is also consistent with the CVF, which views different culture types as complementary rather than competing.

Furthermore, innovation has not historically been highly prioritized within the military, given the especially high emphasis placed on structure and hierarchy[12] and the general unwillingness to take risks.[13] This is not to say that innovation does not occur within the military, but rather that it is not an aspect of DoD culture emphasized as prominently compared with other values. For example, a *Sense of Duty*, which refers to the "degree to which an organization feels a profound obligation and allegiance to support a mission or cause," has been proposed as a unique and critical component of military culture.[14] Military values can be further exemplified in the stated values and creeds of various services of DoD and its branches of service. For example, the U.S. Army Soldier's Creed states, among other things, "I will always place the mission first,"[15] which emphasizes such values as selflessness and duty. Likewise, the U.S. Navy Sailor's Creed states, among other things, "I will obey the orders of those appointed over me," which emphasizes such values as order and obedience.[16] Although

---

bell, A. Kim, K. Finney, K. Barr, and A. M. Hassan, "A Brief Introduction to the Military Workplace Culture," *Work*, Vol. 50, No. 1, 2015. It should be noted, however, that the term *greedy institutions* is a better descriptor of the service members within DoD rather than the civilian workforce.

[10] J. E. Coll, E. L. Weiss, P. Draves, and D. Dyer, "The Impact of Military Cultural Awareness, Experience, Attitudes, and Education on Clinician Self-Efficacy in the Treatment of Veterans," *Journal of International Continuing Social Work Education*, Vol. 15, No. 1, January 2012.

[11] B. Robers, "Public Understanding of the Profession of Arms," *Military Review*, November–December 2012.

[12] J. K. Tinoco and A. Arnaud, "The Transfer of Military Culture to Private Sector Organizations: A Sense of Duty Emerges," *Journal of Organizational Culture, Communications and Conflict*, Vol. 17, No. 2, 2013.

[13] B. W. Everstine, "Esper: Culture Change in DoD Needed to Improve Acquisition Process," *Air Force Magazine*, January 2020.

[14] Tinoco and Arnaud, 2013.

[15] U.S. Army, "Soldier's Creed," webpage, undated.

[16] U.S. Army, undated.

subcultures (i.e., groups whose cultures have slightly discernable beliefs and values relative to the larger organization's beliefs and values) certainly can exist both between and within the various branches of service, DoD and its various branches are united by common values, such as integrity, honor, loyalty, and duty.[17] Thus, it is possible to talk about the dominant U.S. military culture while also acknowledging that subcultures do indeed exist, which may conflict with each other. Military culture is also unique in that its organizational culture is continually indoctrinated within its members via a rigorous socialization process and, therefore, is strongly shared and held by its members. As noted by Meyer, McCarroll, and Ursano, "Good or bad, indoctrination into military culture is so profound that it can fundamentally change a worldview, often impeding transition back to civilian life."[18] Although culture is capable of influencing people's perceptions in typical organizational settings (i.e., nonmilitary organization), the culture of military organizations tends to influence these views to much greater extents (i.e., is more impactful and salient).[19]

In contrast to DoD, the mission, values, competitive environment, and organizational culture of U.S. technology firms are quite different. Indeed, U.S. technology firms tend to emphasize such values as risk, autonomy, innovation, profits, growth, and a disdain toward rules and hierarchy.[20] One study designed to compare companies in a variety of industries along nine dimensions of organizational culture found several notable differences between the Tech Giants category and the Aerospace and Defense category (the closest proxy in the study for DoD).[21] For instance, of the companies examined, technology organizations scored higher on innovation (i.e., creating new products, technologies, work methods) and agility (i.e., reacting more quickly to changes in the market and work environment) than defense companies on average.

Like DoD and its various branches of service, technology firms also possess a set of stated values and philosophies that presumably exemplify their organizational culture. For example, Netflix developed a culture deck (i.e., detailed slideshow) that lists its guiding values, which include such factors as innovation, curiosity, and courage.[22] Likewise, Google espouses

---

[17] K. Jackson, K. L. Kidder, S. Mann, W. H. Waggy II, N. Lander, and S. R. Zimmerman, *Raising the Flag: Implications of U.S. Military Approaches to General and Flag Officer Development*, Santa Monica, Calif.: RAND Corporation, RR-4347-OSD, 2020; and J. M. Mattox, "Values Statements and the Profession of Arms: A Reevaluation," *Joint Force Quarterly*, Issue 68, First Quarter 2013.

[18] E. G. Meyer, J. E. McCarroll, and R. J. Ursano, eds., *U.S. Army Culture: An Introduction for Behavioral Health Researchers*, Bethesda, Md.: Center for the Study of Traumatic Stress, 2017.

[19] Ployhart, Hale, and Campion, 2014.

[20] A. Grinstein and A. Goldman, "Characterizing the Technology Firm: An Exploratory Study," *Research Policy*, Vol. 35, No. 1, February 2006.

[21] D. Sull, C. Sull, and A. Chamberlain, *Measuring Culture in Leading Companies*, Cambridge, Mass.: MIT Sloan Management Review, 2019.

[22] Netflix, undated.

such values and principles as "fast is better than slow" and "great just isn't good enough."[23] To illustrate how the stated values of U.S. technology firms compare with the stated values of DoD and its branches of service, the values of five prominent technology companies (Amazon, Facebook, Google, Microsoft, and Netflix) and the values of four of DoD's service branches (Army, Air Force, Marine Corps, Navy) are presented in Table 1.1. The U.S. military frequently emphasizes such values as courage, commitment, and honor, while Silicon Valley tends to emphasize such values as innovation, taking quick action, and being customer-focused. In many ways, the values espoused by these different industries are divergent. For example, Amazon encourages leaders to "have a backbone; disagree and commit," even when doing so is uncomfortable. In the military, however, loyalty and duty are highly valued.[24] Thus, disagreeing with others (e.g., military officers), especially in a highly hierarchical organization, may not be as tolerated because this is not congruent with the dominant cultural values of the military.[25] In some instances, however, the values shared across industries are aligned. For example, integrity and courage are common across industries (e.g., DoD, Army, and Netflix).

Because leadership plays a crucial role in shaping a company's culture during its founding period, statements made by company leaders and founders can also provide important insights about the culture of these organizations.[26] For example, when Google was founded by Sergey Brin and Larry Page, the founders stated, "Our management philosophy amplified that quality employees who are motivated do not need to be managed."[27] Such a philosophy no doubt helped solidify the importance of autonomy and freedom as an important aspect of Google's culture. As another example, in the first Amazon letter to shareholders, Jeff Bezos wrote an entire section entitled, "Obsess Over Customers," which exemplifies the company's external focus and solidified its emphasis on the customer experience.[28] To sum, although there are clearly differences in the cultures of DoD and Silicon Valley, it is important to acknowledge that there are also commonalities in the cultures of these two types of organizations (e.g., both Netflix and DoD include courage as a stated value that is important to their organizational cultures). However, the manner in which such differences manifest (i.e., their artifacts) is probably somewhat variable across the two. Furthermore, such similarities and differences are likely a matter of degree rather than being truly dichotomous. For instance,

---

[23] Google, undated.

[24] Coll et al., 2012.

[25] A. Milburn, "Losing Small Wars: Why US Military Culture Leads to Defeat," *Small Wars Journal*, September 12, 2021.

[26] Schein, 2010.

[27] T. A. Finkle, "Corporate Entrepreneurship and Innovation in Silicon Valley: The Case of Google, Inc.," *Entrepreneurship Theory and Practice*, Vol. 36, No. 4, July 2012.

[28] J. P. Bezos, "Letter to Shareholders," AboutAmazon.com, 1997.

it is probably more accurate to describe Silicon Valley as valuing innovation more than DoD and not as innovative while DoD is not innovative.

Mapping Department of Defense and Silicon Valley Culture on the Competing Values Framework

Because an organization's culture can be described by a seemingly infinite number of dimensions, having a standardized framework to compare cultures is helpful for simplifying these comparisons and making sense of culture. As noted earlier, we relied primarily on the CVF to describe the differences between DoD and Silicon Valley.[29] Because the CVF was initially created to describe a typical (i.e., nonmilitary) organization, it is possible that CVF's four classifications fail to capture important components of military organizational culture. Therefore, we also rely on the Sense of Duty dimension identified by Tinoco and Arnaud to further compare DoD and Silicon Valley and ensure a comprehensive assessment of culture.[30] Using the CVF and the Sense of Duty dimension, we propose five hypotheses about the differences in organizational culture between our chosen groups.

Hypothesis 1: Department of Defense Is More Likely to Be Characterized as a Hierarchy Culture Compared with Silicon Valley

The presence of structured processes, policies, and regulations throughout the U.S. military, as well as the strong emphasis placed on the chain of command, support the notion that DoD might be expected to exhibit characteristics of a Hierarchy culture (i.e., "do things right").[31] The importance of rules as well as the value placed on stability are typical of other organizations that display more traits of a Hierarchy culture compared with the other dimensions of organizational culture.[32] Indeed, there is evidence that the U.S. military views its culture as very rule-based and policy-focused, despite a desire by its leadership to change the culture of the military to become more innovative and flexible.[33] Further support for this initial supposition comes from studies that have used the CVF to assess specific components of the U.S. military. For example, one examination of Air Force units found that Hierarchy frequently

---

[29] Tinoco and Arnaud, 2013.

[30] Tinoco and Arnaud, 2013.

[31] S. J. Gerras, L. Wong, and C. D. Allen, *Organizational Culture: Applying a Hybrid Model to the U.S. Army*, Carlisle Barracks, Pa.: U.S. Army War College, 2008.

[32] A. Pollman, "Framing Marine Corps Culture," *Proceedings*, June 2018; and T. M. Williams, "Practicing What We Preach: Creating a Culture to Support Mission Command," *Small Wars Journal*, blog post, July 2019.

[33] J. G. Pierce, *Is the Organizational Culture of the U.S. Army Congruent with the Professional Development of its Senior Level Officer Corps?* Carlisle Barracks, Pa.: U.S. Army War College, 2010.

emerged as one of the most prominent dimensions.[34] Likewise, a study focusing on Marine Corps officers also found that Hierarchy was often one of the most important cultural dimensions to emerge.[35] Interestingly, both these studies also found that the military units that they examined were frequently characterized as a Market culture. Given that Market cultures focus on aggression and competition (while still emphasizing stability and control), it makes sense that the military's organizational culture might have a relatively greater degree of affinity with the traits of a Market culture.

The emphasis placed on hierarchy contrasts with Silicon Valley, where software engineers often maintain a disdain toward bureaucratic rules.[36] The greater emphasis placed on flexibility and freedom suggests that software firms are much less likely to value adherence to regulations and restrictions on their autonomy. Furthermore, the goal of many technology firms to "disrupt" established business models and assumptions about work indicate that there is a high premium placed on risk and uncertainty. This contrasts with DoD, where stability and order are highly valued. Although both DoD and Silicon Valley are likely to be characterized as hierarchical cultures to some degree (e.g., it is unlikely that Silicon Valley companies reject all rules and regulations), we predict that this culture dimension will be more prominent for DoD.

Hypothesis 2: Silicon Valley Is More Likely to Be Characterized as an Adhocracy Culture Compared with the Department of Defense

In many ways, the primary mission of technology firms is to innovate.[37] This is in contrast to the primary mission of DoD, which is to create military forces prepared to engage in warfare if needed to protect the security of the United States.[38] Furthermore, many of the stated values of American technology firms emphasize values that are indicative of Adhocracy cultures (i.e., "do things first"—e.g., focus on innovation, creativity, risk). For DoD, however, its stated values tend to emphasize such concepts as integrity, honor, and courage that are not as prominent within Adhocracy cultures. This is not to say that there are not aspects of Adhocracy cultures that are valued by DoD, but that DoD has values that seem more likely to align with the Hierarchy dimension of culture. Thus, taken as a whole, we predict that the Adhocracy dimension will be more prominent for Silicon Valley than DoD.

---

[34] R. Erhardt, *Cultural Analysis of Organizational Development Units: A Comprehensive Approach Based on the Competing Values Framework*, dissertation, Atlanta, Ga.: Georgia State University, 2018.

[35] A. Pollman, *Diagnosis and Analysis of Marine Corps Organizational Culture*, Executive Master of Business Administration Capstone Project Report, Monterey, Calif.: Naval Postgraduate School, March 2015.

[36] V. Khosla, "The Silicon Valley Culture," *Medium*, January 17, 2018; and A. Pardes, "Silicon Valley Ruined Work Culture," *Wired*, February 24, 2020.

[37] Grinstein and Goldman, 2006.

[38] J. Mattis, *Summary of the 2018 National Defense Strategy of the United States of America*, Washington, D.C.: U.S. Department of Defense, 2018.

Hypothesis 3: Silicon Valley Is More Likely to Be Characterized as a Market Culture Compared with the Department of Defense

Corporations often put a focus on growth and profits at the heart of everything they do. This is reflected in several of the stated values listed by the technology firms we analyze in this study, such as Amazon's stated focus on "customer obsession" or Google's assertion that "Great just isn't good enough." Although DoD might also be expected to display some degree of focus on competition with its rival organizations, our initial prediction is that the Market dimension of culture (i.e., "do things fast") will be more prominent for the Silicon Valley companies when compared with DoD.

Hypothesis 4: Silicon Valley and Department of Defense Will Rank Similarly When Measured on the Clan Dimension of Organizational Culture

There are reasons to believe that both the military and technology firms might value the traits characterized by the Clan culture. Both organizations typically emphasize teamwork and collaboration among their members to achieve shared goals. As one study on the military notes, "Teams are the foundational building blocks of the military,"[39] while a similar look at how Silicon Valley companies operate explains, "It's a culture where teams self-organize; people from various functions come together to work on specific projects by habit, not by exception; and good ideas gain momentum organically by attracting talent from around the business."[40] Consequently, we predict that both types of organization might exhibit traits of a culture that emphasizes "doing things together," as Clan cultures typically do.

Hypothesis 5: Department of Defense Is More Likely to Be Characterized by a Sense of Duty Culture Compared with Silicon Valley

Lastly, for our fifth hypothesis, the Sense of Duty dimension,[41] was developed specifically for the U.S. military because of the deficiencies in extant theoretical frameworks.

This is consistent with the idea that the military would be much more likely to emphasize such values as integrity and honor compared with U.S. technology firms (see Table 2.1). Figure 2.1 provides our initial hypotheses about how the cultures of DoD and Silicon Valley might map onto the CVF.

---

[39]   G. F. Goodwin, N. Blacksmith, and M. R. Coats, "The Science of Teams in the Military: Contributions From Over 60 Years of Research," *American Psychologist*, Vol. 73, No. 4, 2018.

[40]   H. Martins, Y. B. Dias, and S. Khanna, "What Makes Silicon Valley Companies so Successful," *Harvard Business Review*, April 26, 2016.

[41]   Tinoco and Arnaud, 2013.

FIGURE 2.1

Hypothesis of How the Department of Defense's and Silicon Valley's Organizational Cultures Might Map onto the Competing Values Framework

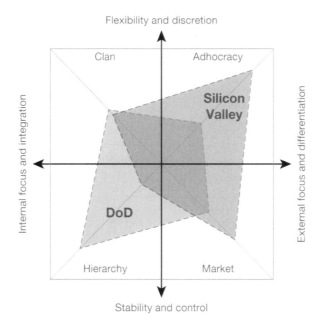

NOTE: In this conceptual mapping, the axes represent the prominence of each cultural dimension. For example, we expect a large difference between DoD and Silicon Valley on the Hierarchy cultural dimension (with DoD being much higher), although a much smaller difference is expected for the Clan dimension. Silicon Valley appears larger because of its positioning on the axis and not for any substantive reasons.

CHAPTER THREE

Method

To quantitatively compare the organizational cultures of DoD and Silicon Valley, we employed a text analytics/mining approach using RAND-Lex (described in detail below). Because language serves as a mechanism by which people articulate and espouse values, the words that people use to talk and write about DoD and Silicon Valley (e.g., military leader speech transcripts, shareholder letters, news stories, mission statements) may be used to make inferences about the cultures of these organizations.[1] Indeed, research suggests that text-mining and natural language processing techniques represent viable strategies for assessing organizational phenomena, including organizational culture specifically.[2] A summary of the method we employed and some of the specific components of each step can be seen in Figure 3.1.

The first step of this process was to identify a set of documents that would contain insights about the organizational cultures of DoD and Silicon Valley. To determine which documents should be targeted for the corpus, various informal discussions with subject-matter experts (e.g., fellow researchers, military personnel) took place. Using the information from such discussions, as well as a review of the literature on organizational culture and text analytic methodology, we compiled a list of viable sources that would presumably contain insights about the cultures of DoD and Silicon Valley. This included news stories, magazine articles, blog posts, interview transcripts, speeches, shareholder letters, book summaries, military manuals, organization assessment reports, culture decks, and stated values. The following EBSCO and Proquest databases were used to search for the information: business book summaries, business sources, military collections, Nexis Uni, and U.S. major dailies. Other sources of information about organizational culture included Air University Press, the Defense Technical Information Center, National Defense University Press, Strategic Studies Institute, specific company websites (e.g., for company values and shareholder letters), specific defense publication websites (e.g., for military values and creeds), and general web searches. Targeted searching was also used to acquire specific information that was not acquired from the above procedures. For example, *Forbes* and *Entrepreneur* magazines were targeted to locate

---

[1] S. Pinker, *The Stuff of Thought: Language as a Window into Human Nature*, New York: Viking, 2007; and Y. R. Tausczik and J. W. Pennebaker, "The Psychological Meaning of Words: LIWC and Computerized Text Analysis Methods," *Journal of Language and Social Psychology*, Vol. 29, No. 1, 2010.

[2] S. Pandey and S. K. Pandey, "Applying Natural Language Processing Capabilities in Computerized Textual Analysis to Measure Organizational Culture," *Organizational Research Methods*, Vol. 22, No. 3, 2019.

FIGURE 3.1
Summary of Project Approach

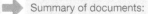

Specify corpus	Compile documents	Clean documents
• Literature review, conversations with subject-matter experts (RAND researchers, military personnel, mentors)	• Many RAND databases were examined (military collection, U.S. major dailies, Nexis Uni)	• Convert all pdfs to txt files (manually or via ghostscript)
• Focus on leadership (not rank-and-file employees)	• Other sources were also examined (company/military webpages, web searches)	• Remove artifacts (i.e., metadata from sources; document URLs, publisher information, email subscription requests, page numbers, etc.)
• Sources need to contain insights about organizational culture	• Targeted searches also used to find specific information (*Forbes*, *War on the Rocks*)	Summary of documents:
• News stories, speeches, stated values, interview transcripts, shareholder letters, military manuals, book summaries, organizational assessment reports, culture decks	• DoD branches and specific technology companies examined	– **522 total documents** – **209 DoD documents** – **313 Silicon Valley documents**
	• Equivalent information sources (shareholder letters/DoD organizational assessment reports)	– **743,466 total words** – **372,117 DoD words** – **371,349 Silicon Valley words**
	• Search terms: "organizational culture," "corporate culture," among others.	

articles about Silicon Valley culture, and *War on the Rocks, Air Force Magazine, Small Wars Journal, Foreign Policy*, and *Task & Purpose* were all targeted to locate articles on military culture. Generally, we limited our searches to information from the past ten years. In some cases, exceptions were made for particularly important documents (e.g., documents written during a company's founding period). Some examples of search terms that were used include "organizational culture," "military culture," "Silicon Valley culture," "Department of Defense values," "corporate values," and other similar variations. Search terms were broad and did not focus on specific culture dimensions to ensure that the results would not be biased.

In general, we attempted to locate equivalent sources of information about organizational culture for both DoD and Silicon Valley. The rationale for this decision was to mitigate potential issues with comparing nonequivalent information. For example, although we were able to locate shareholder letters for Amazon, Google, and Microsoft, there is no exact military equivalent of this information. As a proxy for this type of source, however, we compiled some of the recent DoD organizational assessment reports because these reports are intended to document the state of DoD (similar to how shareholder letters document the state of the company).[3] Likewise, although we were able to locate various speech transcripts for military leaders, interview transcripts tended to be more common for Silicon Valley leaders. Because

---

[3] Office of the Director of Administration and Management, U.S. Department of Defense, "Organizational Assessment," webpage, undated.

both of these speech mediums provide leaders an opportunity to articulate information about their organization's values and culture, we considered them to serve a similar purpose. There will always be some degree of difference between a corpus of documents describing a government agency, such as DoD, and a private sector business, because these organizations have different stakeholders, operate within a different context, and ultimately have a different purpose. However, the accumulated documents provide the best available look at the culture that the leaders of each organization wish to inculcate into their domains.

Although the corpora for DoD and Silicon Valley were not exactly equivalent, our efforts to ensure general equivalence can provide some assurance that the results of the RAND-Lex analysis are meaningful. We also attempted to ensure that the documents that we gathered were relevant to culture. This was accomplished by reviewing a random subset of the documents and using our judgment to determine the appropriateness of the content. There were inherent scalability issues with the procedure that we used that required various methodological trade-offs. For instance, although reading every document and determining its appropriateness (probably with the use of multiple raters) would be optimal, such a process would be too time-consuming (and is rarely done within text-analytics frameworks), especially when thousands of potential documents need to be reviewed. The purpose of this additional validity check was to provide assurance that our document selection method was deriving appropriate information (i.e., documents relevant to organizational culture) for the corpus. The random sampling/review of documents indicated that this was the case and thus served as a useful additional screening method to ensure rigor in this process.

Once the final set of documents was compiled, we then cleaned the text of artifacts (i.e., superfluous information that could bias the results).[4] This entailed removing all of the metadata that were not directly relevant to the primary text and included such information as page numbers, publication dates, subscription requests, and output generated by the various databases that were used (e.g., source type, document URL, publisher). This effort was accomplished by converting the pdf files to txt files (either manually or via a ghostscript code) and then manually cleaning the documents. Although this process was not perfect, it did permit us to substantially reduce the number of artifacts present in each of the documents.[5] This process of gathering and cleaning documents yielded a set of 522 total documents (DoD number of documents = 209; Silicon Valley number of documents = 313). Of the 209 DoD documents, 144 were news stories, 42 were speeches by military leaders, and 23 were other documents (e.g., DoD organizational assessment reports, stated values and creeds, relevant sections of military manuals). Of the 313 Silicon Valley documents, 243 were news stories, 34 were company leader interviews, and 36 were other documents (e.g., shareholder letters,

---

[4]   V. B. Kobayashi, S. T. Mol, H. A. Berkers, G. Kismihók, and D. N. Den Hartog, "Text Mining in Organizational Research," *Organizational Research Methods*, Vol. 21, No. 3, July 2018.

[5]   J. Kavanagh, W. Marcellino, J. S. Blake, S. Smith, S. Davenport, and M. Gizaw, *News in a Digital Age: Comparing the Presentation of News Information over Time and Across Media Platforms*, Santa Monica, Calif.: RAND Corporation, RR-2960-RC, 2019.

stated values, culture decks, book summaries). The entire corpus of documents contained 743,466 words. The corpus of DoD documents consisted of 372,117 total words, while the corpus of Silicon Valley documents consisted of 371,349 total words. Although not exhaustive, the goal of this process was to gather a sufficiently representative corpus of documents to make inferences about the organizational cultures of DoD and Silicon Valley.

CHAPTER FOUR

Analyses

To quantitatively compare the organizational cultures of DoD and Silicon Valley on the basis of the documents that we compiled, we used RAND-Lex. RAND-Lex is a proprietary text analytics tool that combines machine learning and qualitative analyses to analyze corpora of documents.[1] Some specific analyses that can be performed with RAND-Lex that are relevant to this project include keyness testing/analysis, collocate analysis, and stance comparison analysis.

Keyness Testing

Keyness testing represents an empirical strategy for determining whether certain words are noticeably overpresent or underpresent in a collection of documents relative to a baseline collection of documents. To compare the corpus of DoD and Silicon Valley documents, we began by conducting a series of keyness tests on a set of predefined words. The predefined words that we examined were selected so as to be representative of the five cultural dimensions of interest (Hierarchy, Adhocracy, Clan, Market, and Sense of Duty). Keywords for CVF dimensions were derived from relevant academic sources and from the authors who developed the CVF to ensure a comprehensive identification of keywords.[2] Keywords for the Sense of Duty dimension were also derived from academic sources.[3] A summary of the keywords, including variations thereof, for each dimension can be seen in Table 4.1. The Hierarchy dimension consisted of 55 words, the Adhocracy dimension consisted of 68 words, the Clan dimension consisted of 66 words, the Market dimension consisted of 67 words, and the Sense of Duty dimension consisted of 41 words.

Next, a series of keyness tests were conducted on every word identified for each of the five dimensions of organizational culture (297 total comparisons). To accomplish this, DoD documents were specified as the target group, while the Silicon Valley documents were specified as the baseline group.[4] Thus, words that are overpresent appeared more frequently in

---

[1] D. Irving, "Big Data, Big Questions," *RAND Review*, October 16, 2017.

[2] Cameron and Quinn, 2006; Cameron, 2009; Erhardt, 2018; Hartnell, Ou, and Kinicki, 2011.

[3] Tinoco and Arnaud, 2013.

[4] We acknowledge that the culture of organizations can change over time. However, the corpus of documents was not large enough to stratify the results by year. Exploring a bottom-up view of culture might yield larger data sets that could be stratified in this way.

TABLE 4.1

Complete List of Organizational Culture Keywords

Culture Dimension	List of Keywords
Hierarchy	accountability, accountable, authority, authorities, bureaucracy, bureaucratic, capable, capability, capabilities, communicate, communicators, communicates, communications, communications, communicating, consistent, consistently, control, controlling, controlled, coordination, coordinate, efficiency, efficiencies, efficient, efficiently, formal, formally, hierarchy, hierarchical, monitor, monitored, organization, organizations, organization's, organizational, policy, policies, predict, predictability, predictable, procedures, rank, ranks, regulations, regulation, reliable, rules, rule, stability, stable, standard, standards, structure, structured
Adhocracy	adaptable, adapt, adaptability, adapted, adapting, adaption, adaptive, agile, agility, anticipate, autonomous, autonomy, change, changed, changes, changing, create, created, creates, creating, creative, creatively, creativity, detail, detailed, details, dynamic, dynamics, entrepreneurial, entrepreneurs, entrepreneurship, experiment, experimental, experimentation, experimenting, flexibility, flexible, free, freedom, freedoms, future, futures, grow, growing, grown, growth, imagine, imagined, innovate, innovating, innovation, innovations, innovative, innovators, new, opportunities, opportunity, research, risk, risks, risky, stimulate, temporarily, uncertain, uncertainties, uncertainty, variety, vision
Clan	Attached, cohesion, cohesive, cohesiveness, collaboration, collaborative, collaborate, competence, competencies, competent, competency, concern, concerns, concerning, concerned, consensus, development, developmental, develop, develops, developing, developed, empower, empowers, empowered, empowering, empowerment, human resources, individual, individuals, individually, involvement, involved, involve, involving, involves, loyalty, loyal, mentor, mentorship, mentoring, mentors, morale, participation, relationship, relationships, satisfaction, self-development, skill, skills, support, supports, supporting, supportive, supported, team, teams, teamwork, teammates, training, trained, train, trains, trust, trusted, voice
Market	achievement, achievable, achieve, achieved, achievements, achieving, aggressively, aggression, aggressive, communicate, communicates, communicating, communications, communications, communicators, competence, competencies, competency, competent, competing, competition, competitions, competitive, competitiveness, competitors, contract, contracting, contractor, contractors, contracts, control, controlled, controlling, coordinate, customer, customers, energy, environment, environments, external, fast, faster, goal, goals, market, markets, perform, performance, performed, performers, performing, plan, planned, planners, planning, plans, productive, productivity, profit, profitable, profits, rapid, rapidly, result, results, return, speed
Sense of Duty	allegiance, authority, cause, conduct, courage, courageous, courageously, duty, duties, devotion, discipline, disciplined, disciplinary, disciplines, honor, honorable, honored, honoring, honors, integrity, loyalty, loyal, mission, obey, purpose, purposes, respect, respectful, respectively, respective, respects, respected, sacrifice, sacrifices, sacrificed, sacrificing, selfless, selflessly, service, subordinate, subordinates

NOTES: Keywords were derived from Cameron and Quinn, 2006; Cameron, 2009; Erhardt, 2018 (as adapted from Müller and Nielson, 2013); and Hartnell et al., 2011, to ensure a comprehensive assessment of the CVF. Keywords for the Sense of Duty dimension were derived from Tinoco and Arnaud, 2013. Words within the dimensions are not necessarily mutually exclusive, although the words included above represent the most important words associated with each dimension. Moreover, including the same words across multiple dimensions is not ideal and can obfuscate results.

the corpus of DoD documents compared with the corpus of Silicon Valley documents. Conversely, words that are underpresent appeared more frequently in the corpus of Silicon Valley documents compared with the corpus of DoD documents. We set a minimum word frequency of 5 so that very infrequently used words would not be included in the analysis.[5] For each comparison, we examined the loglikelihood (*LL*), whether the word was overpresent or underpresent, and the number of times the word appeared (i.e., frequency) within DoD and Silicon Valley documents. Within this context, *LL* is a measure of confidence that the target word is truly overpresent or underpresent. Thus, this statistic indicates whether the target word frequency differences are due to chance. *LL* values greater than 11 represent statistically significant differences in word frequency. Generally, higher *LL* values are indicative of greater differences in word frequency and can therefore be used as way to determine whether the differences in word frequencies are meaningful. The results for every word that we examined for each of the five dimensions are shown in Tables A.1 through A.5 in Appendix A. To better interpret the findings of the keyness testing for the 297 words that were examined, we averaged the above information for each dimension. We also computed some additional information to better understand these findings. For example, we computed the mean *LL* value for words that were overpresent and underpresent. We also calculated the percentage of overpresent and underpresent words with an *LL* value of greater than 11 so as to better understand how many of these differences in word frequencies are actually meaningful. Table 4.2 details the technical results.

Regarding word frequency, the results indicated that words associated with the Hierarchy dimension appeared more frequently in the DoD documents compared with the Silicon Valley documents. Similarly, words associated with the Sense of Duty dimension appeared more frequently in the DoD documents compared with the Silicon Valley documents. This dimension represented the largest difference in word frequencies. Words associated with the Clan dimension also appeared more frequently in the DoD documents compared with the Silicon Valley documents. In contrast, words associated with the Adhocracy dimension appeared more frequently in the Silicon Valley documents compared with the DoD documents.[6] A similar result emerged for the Market dimension in that words appeared more frequently in the Silicon Valley documents compared with the DoD documents. This information is summarized in both Table 4.2 and Figure 4.1.

---

[5] A lower threshold (e.g., 1) would have resulted in infrequently used words being included, which is not ideal. A higher threshold (e.g., 5) would have excluded many meaningful words, which is also not ideal. A threshold of 5, therefore, strikes a useful balance between these two extreme thresholds.

[6] An inspection of Table A.2 reveals that the term *new* for the Adhocracy dimension might represent an outlier, as it was used very frequently across both DoD and Silicon Valley documents relative to the other words within this dimension. Although computing the mean word frequency without this word slightly reduces the Adhocracy word frequency means for DoD ($M = 58.45$) and Silicon Valley ($M = 68.24$), this does not change the rank order of the results very much. Furthermore, the *LL* for the term *new* is 189.65, which suggests that the difference in the frequency of this word is very meaningful. Consequently, we retained the term *new* for the keyness analysis because this word is an integral component of the Adhocracy culture dimension.

TABLE 4.2
Summary of Keyness Analysis Results

Dimension	M LL	M Over LL	M Under LL	Over Words (%)	Over LL > 11 (%)	Under LL > 11 (%)	DoD M Word Freq.	SV M Word Freq.	Word Freq. Dif.
Hierarchy	15.30	20.36	1.77	73	48	0	53.93	33.65	20.28
Adhocracy	22.00	15.16	28.46	49	36	43	67.66	87.47	−19.81
Clan	29.20	31.40	21.69	77	43	47	65.36	43.65	21.71
Market	36.28	20.26	58.59	58	31	43	44.22	66.00	−21.78
Sense of Duty	54.84	54.84	—	100	49	—	84.49	21.07	63.42

NOTES: *LL > 11* represents a statistically significant result. *Words* in the fourth column's header refers to the total number of words examined for the dimension. *Over* refers to words that were conspicuously overpresent (i.e., appeared more) in the target data set (DoD documents) compared with the baseline data set (Silicon Valley documents). *Under* refers to words that were conspicuously underpresent (i.e., appeared less) in the target data set (DoD documents) compared with the baseline data set (Silicon Valley documents). Positive difference values indicate that the word frequency was higher for DoD compared with Silicon Valley. M = Mean. SV = Silicon Valley. Word Freq. = Word Frequency. Word Freq. Dif. = Word Frequency Difference.

To corroborate the above findings, we also examined the percentage of overpresent and underpresent words with an *LL* value of greater than 11 (i.e., instances where significant differences existed). For the Hierarchy dimension, almost half of overpresent words (i.e., those appearing more frequently in the DoD documents) displayed statistically significant results (i.e., had *LL* values of greater than 11), while none of the underpresent words (i.e., those appearing more frequently in the Silicon Valley documents) displayed statistically significant results. Thus, although nearly one-third of words constituting the Hierarchy dictionary appeared more frequently in the Silicon Valley documents, the greater frequency of these words was not meaningful. Likewise, for the Sense of Duty dimension, almost half of the overpresent words displayed statistically significant results. Thus, of the words that appeared more frequently in the DoD documents (which is all of the words in this case), the differences in word frequency were meaningful for nearly half of the words within this category. The opposite pattern was observed for the remaining dimensions. Specifically, for the words that appeared more frequently in the Silicon Valley documents, there was a greater percentage of words that displayed statistically significant results for the Adhocracy, Clan, and Market dimensions compared with words that appeared more in the DoD documents. Thus, for words that appeared more in the Silicon Valley documents for these latter three culture dimensions, the difference in the frequency at which these words appeared was more meaningful compared with the difference in frequency for words that appeared more within the DoD documents. These results are summarized in Figure 4.2.

Overall, the results of these three subanalyses (word frequencies, the mean for *LL*, Over/Under *LL* greater than 11) are consistent, with the exception of the Clan dimension. For example, although the word frequency count and mean *LL* value were higher for the DoD documents for the Clan dimension, the percentage of words that displayed statistically signif-

FIGURE 4.1

Mean Department of Defense and Silicon Valley Word Frequencies Across Organizational Culture Dimensions

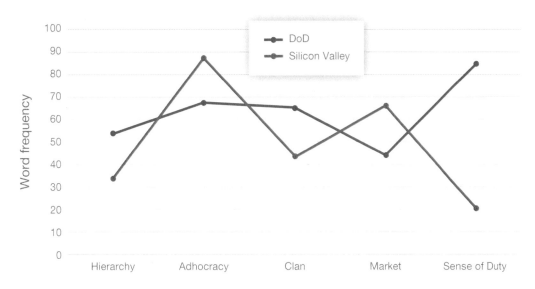

icant results was slightly higher for the Silicon Valley documents. These somewhat ambiguous findings might reflect the fact that this culture dimension is prominent within both the Silicon Valley and DoD documents. For the remaining culture dimensions, the results are clearer. Words associated with the Hierarchy and Sense of Duty dimensions are more prominently used within the DoD documents, while words associated with the Adhocracy and Market dimensions are more prominently used within the Silicon Valley documents.

Collocate Analysis

To supplement the above findings, we conducted a collocate analysis to try to understand how culture is discussed within DoD and Silicon Valley documents. Collocate analysis identifies two-word and three-word sets (i.e., *n*-grams) that frequently co-occur and statistically assesses the meaningfulness of these co-occurrences. For this analysis, we examined the term *culture* and the top 1,000 collocates. In an effort to identify important collocates, we also specified that collocates must occur at least 25 times. This is largely consistent with prior research that has employed RAND-Lex and helps provide assurance that only meaningful collocates are tagged.[7] To assess the strength and meaningfulness (i.e., confidence that the results are not due to chance) of the identified word pairs, we relied on both the pointwise mutual infor-

---

[7]   W. M. Marcellino, K. Cragin, J. Mendelsohn, A. M. Cady, M. Magnuson, and K. Reedy, "Measuring the Popular Resonance of Daesh's Propaganda," *Journal of Strategic Security*, Vol. 10, 2017.

FIGURE 4.2

Percentage of Words with Statistically Significant Differences (i.e., loglikelihood is greater than 11)

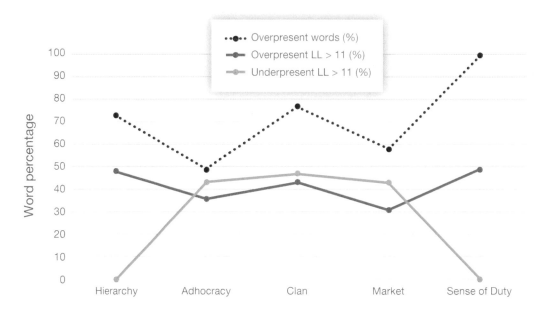

NOTE: *Overpresent* refers to words that appeared more frequently in the DoD corpus compared with the Silicon Valley corpus (represented as a percentage). *Underpresent* refers to words that appeared more frequently in the Silicon Valley corpus compared with the DoD corpus. Thus, *Overpresent* (i.e., dotted line) represents what percentage of words appeared more frequently in the DoD corpus. For example, approximately 70 percent of the words in the Hierarchy dimension appeared more frequently in the DoD corpus.

mation (*PMI*) statistic and Likelihood Ratio (*LR*). *PMI* values greater than 3 indicate a meaningful word pair, and *LR* values greater than 11 indicate that it is unlikely that results are due to chance. Interestingly, for the DoD documents, three of the most meaningful collocates (i.e., those collocates with *PMI* greater than 3 and *LR* greater than 11) were the phrases "cultural change" (*PMI* = 5.19, *LR* = 39.75, Frequency = 30), "change culture" (*PMI* = 3.41, *LR* = 34.61, Frequency = 47), and "culture change" (*PMI* = 3.21, *LR* = 27.62, Frequency = 41). Thus, 118 instances of the term "culture" (and "change") were used to refer to culture change. The phrase "culture of excellence" was also common (*PMI* = 11.70, *LR* = 1083.19, Frequency = 30) and identified as highly meaningful. Lastly, the phrase "organizational culture" was also identified as meaningful (*PMI* = 5.07, *LR* = 74.97, Frequency = 30), though this phrase is descriptive and somewhat less insightful. For the Silicon Valley documents, the most meaningful collocates tended to be descriptive. For example, "corporate culture" (*PMI* = 4.78, *LR* = 83.79, Frequency = 70), "workplace culture" (*PMI* = 4.23, *LR* = 29.15, Frequency = 29), and "company's culture" (*PMI* = 3.63, *LR* = 37.25, Frequency = 46) were some of the top collocates for the Silicon Valley documents. The phrase "create culture" (*PMI* = 3.57, *LR* = 27.60, Frequency = 35) and "culture deck" (*PMI* = 5.47, *LR* = 38.91, Frequency = 27) were also identified as meaningful collocates. Taken as a whole, these results suggest that DoD documents tend to emphasize cultural change when employing the term "culture," but Silicon Valley tends to use this term more descriptively. In other words, DoD may recognize that it needs to change and thus its discussion of culture often occurs in the context of hoping to change it. Silicon Valley tends to be more matter-of-fact in its discussions of culture (i.e., "This is our culture" versus "Here is how we need to change our culture").

Stance Comparison Analysis

To accompany the above findings, we next conducted a stance comparison analysis (SCA). SCA provides a way to statistically compare corpora of documents on 119 predefined language categories consisting of 15 higher categories (e.g., time, emotion, descriptive language, reasoning). An ANOVA Tukey post-hoc test is then used to statistically assess and compare the language categories across the corpora of documents.[8] For example, if one corpus of documents contained more words associated with a particular category (e.g., anger, uncertainty), the ANOVA Tukey post-hoc test would determine whether this difference was significant. A measure of effect size (Cohen's d) is also provided to quantify the meaningfulness of these differences.[9] Although comparisons between language categories could be significant, these differences may be so small as to not be practically significant. The Cohen's d value is thus useful for making determinations about the meaningfulness of an effect.

The goal of this more exploratory analysis was to compare the corpus of DoD and Silicon Valley documents across 119 lower-level language categories (consisting of 15 higher-level categories) and see how people talk about the organizational culture of DoD and Silicon Valley.[10] The stance comparison analysis relies on an a priori taxonomy of linguistic characteristics developed at Carnegie Mellon University.[11] For each category, we examined the results of an ANOVA Tukey HSD post-hoc test and the Cohen's d value (a measure of effective size) to determine which comparisons were meaningful. In general, Cohen's d values from 0.20 to 0.50 are considered small, values from 0.50 to 0.80 are medium, and values greater than 0.80

---

[8] Kavanagh et al., 2019.

[9] The standard equation for Cohen's d is the mean of one group subtracted from the mean of another group divided by the pooled standard deviation:

$$\frac{M_{Group\,1} - M_{Group\,2}}{\sigma}.$$

For more details on Cohen's d, see C. O. Fritz, P. E. Morris, and J. J. Richler, "Effect Size Estimates: Current Use, Calculations, and Interpretation," *Journal of Experimental Psychology: General*, Vol. 141, No. 1, August 2011.

[10] Kavanagh et al., 2019.

[11] For more detailed information, see Carnegie Mellon University, Department of English, "DocuScope: Computer-Aided Rhetorical Analysis," webpage, undated.

FIGURE 4.3

Cohen's *d* Values for Department of Defense Stance Comparison Categories with the Highest Effect Sizes

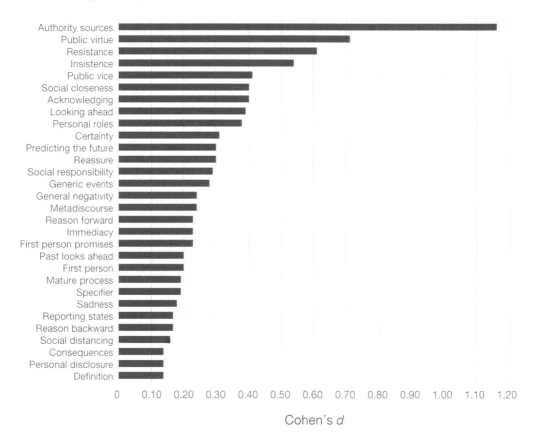

Cohen's *d*

NOTE: Words associated with each category in this figure were used more frequently within the DoD corpus compared with the Silicon Valley corpus. The effect sizes presented in this figure (sorted from largest to smallest) quantify the meaningfulness of these differences such that higher Cohen's *d* values indicate a larger difference in word usage across the categories for the DoD and Silicon Valley corpora.

are large.[12] Of the categories examined, 78 significant differences emerged (the 41 nonsignificant categories can be seen in Appendix B). The Cohen's d values for the comparisons with highest DoD linguistic category scores are presented in Figure 4.3 (M Cohen's d = 0.32), while the Cohen's *d* values for the comparisons with higher Silicon Valley linguistic category scores are presented in Figure 4.4 (M Cohen's d = 0.30).

Taken as a whole, these results suggest that are many differences in the language employed across the DoD and Silicon Valley corpora. For the most part, however, these effect sizes are small and fall somewhere between Cohen's *d* values of 0.20 and 0.50. For DoD, the top

---

[12] J. Cohen, *Statistical Power Analysis for the Behavioral Sciences*, Abingdon, England: Routledge Academic, 1988.

FIGURE 4.4

Cohen's *d* Values for Silicon Valley Stance Comparison Categories with the Highest Effect Sizes

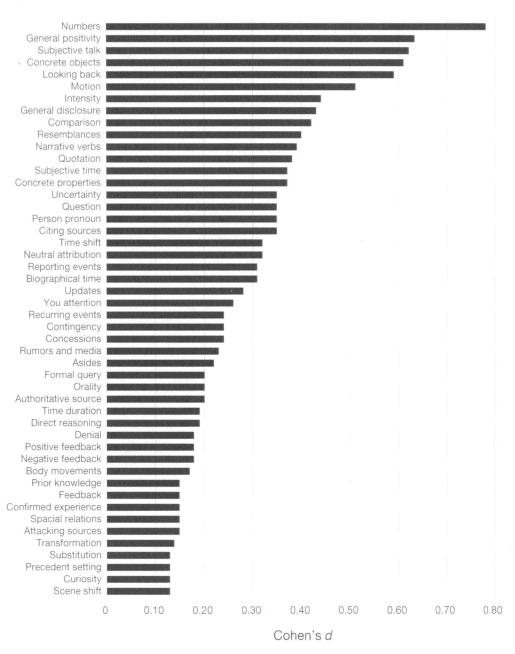

NOTE: Words associated with each category in this figure were used more frequently within the Silicon Valley corpus compared with the DoD corpus. The effect sizes presented in this figure (sorted from largest to smallest) quantify the meaningfulness of these differences such that higher Cohen's *d* values indicate a larger difference in word usage across the categories for the Silicon Valley and DoD corpora.

categories were authority sources (references to respected public or institutional authorities; $d = 1.16$), public virtue (positive, publicly endorsed values, such as justice and fairness; $d = 0.71$), resistance (opposition and/or struggle between ideas, events, groups; $d = 0.61$), insistence (firmness in action and/or reasoning; $d = 0.54$), public vice (behavior and ideas not accepted, such as injustice and unfairness; $d = 0.41$), and social closeness (language that emphasizes belonging and like-mindedness; $d = 0.40$). For Silicon Valley, the top categories were numbers (words that indicate numbers; $d = 0.78$), general positivity (use of positive emotional language; $d = 0.63$), subjective talk (language that acknowledges tentativeness of perceptions; $d = 0.62$), concrete objects (use of concrete objects; $d = 0.61$), looking back (mental leaps into the past; $d = 0.59$), and motion (specific motions, such as walking and jumping; $d = 0.51$). Interestingly, there were no differences in the innovation language category across the DoD and Silicon Valley documents.[13] To illustrate how the stance comparison analysis functions, some examples of the tagged text are provided below. These represent a random selection of two of the highest-scoring SCA sources for the authority sources and innovation language categories. In the first example (see Figure 4.5), words associated with the authority sources language category are tagged (orange) for a random section of a U.S. Army article about culture. In the second example (Figure 4.6), words associated with the innovation language category are tagged (blue) for a random section of a *War on the Rocks* article about culture.

It is interesting that the authority sources language category had the largest effect for DoD. This seems consistent with the fact that authority is an integral part of both Hierarchy and Sense of Duty cultures. According to the above keyness analysis results, these are also two cultures with higher word frequencies for DoD compared with Silicon Valley. It is also interesting that DoD scored high on public virtue and public vice. This seems consistent with the greater word frequency seen for the Sense of Duty dimension for DoD, which emphasizes publicly acceptable values, such as honor, courage, and integrity, and admonishes the absence of such values (e.g., disloyalty, selfishness, lack of integrity). Lastly, social closeness, which represents such notions as belongingness and like-mindedness, seems to be consistent with the fact that words associated with the Clan dimension were frequently used in DoD documents. For Silicon Valley, however, the results are a little more difficult to interpret. In general, the SCA results suggest that the linguistic style that is used to discuss culture within Silicon Valley is somewhat different. For example, numbers, subjective talk, and concrete objects represent some of the largest effects for the Silicon Valley documents. It is, however, interesting to note that more negative language is used within DoD documents, while more

---

[13] To further examine this finding, we conducted the same analysis but excluded *War on the Rocks* articles (25 articles) from the DoD corpus, because these articles are more likely to discuss innovation compared with other sources. Even after removing these articles, we found no differences in the innovation language category. Thus, the innovation language category does not appear to be favored by any particular set of documents in the corpus.

FIGURE 4.5

Stance Comparison Analysis Tagged View for Authority Sources Language Category Words

Trust, Risk and Failure Creating and Sustaining Innovation in Army Culture While the Army is touted as the world's best-trained and best-equipped land force, it must learn and adapt or risk failure in providing national security for an uncertain future. Changing culture and sustaining our competitive advantage will require skilled professionals who know how to promote and protect innovation within the ranks. The core issue may be sustaining innovation in Army organizations, not simply becoming more innovative. Specifically, how should Army leaders address the need for innovation—a notion that inherently conflicts with the larger cultural factors that contribute to the Army's success as a military force? In the professional dialogue on the future of the Army, few topics are discussed more than the need to foster innovation. In November 2014, then-Secretary of Defense Chuck Hagel announced the Defense Innovation Initiative to develop capabilities and capacities for the force of the future. Previous secretaries introduced similar efforts. DoD is consistent in its approach, most recently with the introduction of the Defense Innovation Unit Experimental (DIUx). In practice, innovation can be organizational, including the introduction of a new doctrine, process, or agency such as DIUx; and/or institutional, with an intentional effort to change culture. Through innovation, targeted change may result in the creation of adaptable leaders as well as agile teams and organizations that align to meet the demands of a volatile and uncertain operating environment. Resilient Military Cultures Accordingly, discussions about organizations are fundamentally conversations about culture. Defense critics inside and outside of the profession debate whether the Army can become more innovative. It follows that Army efforts to become more innovative must begin with deliberate introspection of its culture. However, actions to enact desired change are often inhibited by existing practices and structures that are the essence of very resilient military cultures across DoD. One can easily envision an out-brief session of an Army conference. Briefers present their PowerPoint presentations in a prescribed format, with a specific number of slides and a time limit to discuss creative approaches to strategic issues. Of course, the

NOTES: Words highlighted represent the words associated with the authority sources language category. This screenshot is meant to be illustrative and show how RAND-Lex tags the language category words. It is not meant to be exhaustive.

FIGURE 4.6

Stance Comparison Analysis Tagged View for Innovation Language Category Words

acceleration of innovation's pace to remain effective. Unfortunately, a military culture of rigid hierarchy and standardization poses an institutional hurdle that clashes with disruptive thought processes required to develop the kind of innovative capacity recently described by the vice chairman of the Joint Chiefs. John Kotter, professor emeritus at Harvard Business School, explains that strict hierarchies and standardization are inherently risk-averse and resistant to change: Part of the problem is political: Managers are loath to take chances without permission from superiors. Part of the problem is cultural: People cling to their habits and fear loss of power and stature—two essential elements of hierarchies. And part of the problem is that all hierarchies, with their specialized units, rules, and optimized processes, crave stability and default to doing what they already know how to do. Andrew Hill describes in Parameters that militaries otfen see innovation as subversion of the standardization in tools, training, methods, and organization it depends on. As Gen. McChrystal points out, great innovations generally do not come from strict "top-down" organizations. Department of Defense efforts to leverage innovation have largely sidestepped this substantial organizational challenge, instead concentrating on the search for "game-changing" technological advances that tend to come from outside the ranks of the military. Senior defense leaders push development of oiffces like the Air Force's Oiffce of Transformational Innovation and the Department of Defense's Defense Innovation Unit Experimental, which focus on quickly ifnding and integrating disruptive technologies. But, the ability to truly maneuver against unforeseen future challenges requires a more diffuse capacity to innovate across the force. The Solution If the military is determined to "aggressively pursue a path toward institutional strategic agility", as described in the Air Force's Call to the Future, then it must also grow its innovative spirit at the unit level where it is best suited to build broad organizational

NOTES: Words highlighted represent the words associated with the innovation language category. This screenshot is meant to be illustrative and show how RAND-Lex tags the language category words. It is not meant to be exhaustive.

positive language is used in the Silicon Valley documents. The implications of all of these findings are further discussed in the next chapter.

CHAPTER FIVE

Analytic Takeaways

First, our hypothesis was that DoD would exhibit more properties of a Sense of Duty culture (a culture that emphasizes such values as integrity, honor, and courage) compared with Silicon Valley. In line with this prediction, the results demonstrated that DoD and Silicon Valley are furthest apart on this culture dimension. Every word constituting this culture dimension was used more frequently in the DoD corpus of documents constituting with the Silicon Valley documents (Figures 4.1 and 4.2). This factor suggests that the Sense of Duty culture type is quite negligible within Silicon Valley.

However, this could be due, in part, to two mitigating factors. First, this study considered culture from a top-down (i.e., leader-focused), rather than bottom-up, approach. Although corporate CEOs may not seek to promulgate these values, evidence suggests that lower-level employees may be more motivated by a need for purpose and a desire to contribute to something larger than themselves.[1] Second, it is unclear whether the Sense of Duty concept is different for employees in the private sector compared with members of the military. Exploring how members of Silicon Valley's workforce conceptualize their role in society might find a basis for a greater degree of common ground here. Nonetheless, these results suggest that when a top-down approach is taken, Sense of Duty is very important within the U.S. military but far less so within Silicon Valley.

Second, our hypothesis was that DoD would also exhibit more properties of a Hierarchy culture (a culture characterized by order, rules, and structure). The results indicated that DoD and Silicon Valley are also far apart on this dimension of organizational culture. Specifically, words associated with this dimension were used more frequently within the corpus of DoD documents compared with the Silicon Valley documents (Figure 4). As with the Sense of Duty dimension, this suggests that Hierarchy is much less important within U.S. technology firms compared with the U.S. military.

Third, our hypothesis was that Silicon Valley would exhibit more properties of an Adhocracy culture (a culture characterized by autonomy, growth, and innovation). The results documented a notable degree of convergence on the Adhocracy culture dimension. Although Silicon Valley was found to emphasize these traits to a greater degree than DoD, the total

---

[1] Afdhel Aziz, "The Power of Purpose: The Business Case for Purpose," *Forbes*, March 7, 2020.

gap between the two cultures is substantially narrower here compared with the Hierarchy or Sense of Duty dimensions.

However, the results also show that DoD and Silicon Valley view Adhocracy characteristics differently. DoD tends to discuss how it needs to change its culture to emphasize different values than it does today. In contrast, Silicon Valley tends to talk about continuous change and adaption as cultural traits that it has already succeeded in establishing. This finding seems to suggest that the words associated with this dimension are being used by DoD leaders to describe a culture that they hope to achieve rather than the culture as it currently is; further research on how rank-and-file individuals in each culture experience it would help confirm or deny this observation. All in all, the closer alignment between DoD and Silicon Valley in the Adhocracy culture dimension does suggest a potentially useful point of commonality that could be leveraged to improve understanding and communication between the U.S. military and private sector technology firms.

Fourth, our hypothesis was that Silicon Valley would exhibit more properties of a Market culture (a culture that values achievement, competition, and productivity). However, the results showed that Silicon Valley emphasizes the Market culture type more than DoD—but they also showed a relatively small gap between the two communities. Just as in the Adhocracy dimension, Market cultures are valued by both DoD and Silicon Valley, but slightly more by Silicon Valley. This convergence might represent a successful adaption of DoD leadership in discovering how to speak the language of corporate cultures with significant differences from its own.

Lastly, our hypothesis was that Silicon Valley and DoD would display similar affinities for the Clan organizational culture (i.e., a culture that emphasizes such values as development, teamwork, and collaboration). The results confirmed that military and Silicon Valley organizational cultures had the closest alignment in this culture type. Although DoD used words indicative of the Clan culture slightly more often overall, the gap was relatively negligible and was the smallest across the different dimensions of culture. Consequently, appealing to shared values along this dimension might offer the greatest possibility for successful collaborations.

Conclusion

We sought to map and understand the differences in organizational culture between the large technology firms of Silicon Valley and DoD. Although other studies have compared differences in organizational culture between companies in different sectors of the economy, this study is the first to leverage these methods to compare DoD and some of the technology companies leading the United States' investment in the development of AI. As organizations of all kinds come to rely more and more on software applications and AI to transform their operations and organizational efficiency, understanding and bridging any gaps in organizational culture between the technical staff implementing these solutions and the employees of the transformed organization could help ensure that these transformations are more comprehensive and more successful. Bridging these cultural gaps could also potentially help improve communications and enhance partnerships between DoD and Silicon Valley and enable these two communities to better leverage each other's strengths and talents in mutually beneficial ways.

Additionally, this study is one of the first to use quantitative methods and natural language processing (NLP) to measure the organizational culture of either Silicon Valley or DoD along these five dimensions of organizational culture. Although previous studies have primarily relied on a qualitative understanding of the military's values and characteristics to describe its organizational culture, quantitative analytical techniques, such as NLP, can detect unexpected influences on military organizational culture or correct for biases and prejudices that might otherwise go unchallenged. Our analytic finding that Silicon Valley and DoD have only a relatively small gap along the Adhocracy and Market culture dimensions demonstrates how this approach may identify unexpected points of commonality.

Finally, we created NLP dictionaries for each of the five dimensions of organizational culture considered. One benefit of these novel dictionaries is that they provide a standardized framework for examining these specific dimensions of organizational culture. For example, future projects that employ these dictionaries would be able to directly compare their results with those reported here and with any other study that also used these dictionaries. Having a standardized tool like this could make it easier to compare results with subsequent studies. We hope that these dictionaries can be leveraged and further refined by other researchers in future research studies.

Relevance to the Department of Defense

We believe that mapping and understanding how the organizational culture of DoD relates to the organizational culture of technology companies could be interesting to DoD in two ways. First, understanding organizational culture can improve an organization's ability to attract and retain talent. As the research underlying the ASA model has demonstrated, applicants are more likely to be attracted to organizations whose values align with their own, and employees who do not feel their workplace culture is a good fit are much more likely to seek other opportunities. Because many AI experts currently work for U.S. technology companies and are accustomed to their organizational culture, understanding where this organizational culture maps in comparison with DoD's could help DoD expand the range of AI talent available to it.

Additionally, we believe that this type of analysis could help leaders in DoD identify change agents within the military. Senior DoD leaders have expressed their desire to make the military more innovative and more agile. As one example, the current Chief of Staff for the Air Force has headlined his strategic guidance to his service with the motto "Accelerate Change or Lose."[2] Similarly, the military has founded numerous organizations within the department, such as Army Futures Command, the Defense Innovation Unit, and the Strategic Capabilities Office, with a mandate to improve DoD's adoption of advanced technologies and accelerate its pace of innovation. Although the Hierarchy and Sense of Duty cultures are most commonly associated with the military, our analysis indicates that other cultural dimensions, such as the Adhocracy culture (i.e., "get things done"), may have had a greater impact on DoD's organizational culture than previous research has recognized. Identifying suborganizations within DoD that promote concepts associated with this organizational culture or identifying individual officers and other DoD personnel who embrace these concepts could help DoD place individuals more compatible with an Adhocracy culture in the DoD suborganizations intended to promote innovation and agility.

Future Directions

Although this project yielded many relevant findings, it is important to acknowledge the limitations of this research. Here, we note some of these limitations and elaborate on some additional areas for future research beyond what was discussed above. First, the generalizability of these findings is somewhat limited by the nature of the documents that were examined. In particular, the documents represented a top-down perspective on the organizations studied (i.e., they focused more on leadership than rank-and-file employees). Accordingly, it would be beneficial for future studies to take a bottom-up approach (e.g., via focus groups or interviews) that would discover the lived experience of personnel within these organizations and see whether and how that experience differs from the cultural traits that leaders have attempted to promulgate in these organizations. This could be accomplished with Glassdoor.com reviews, employee emails, or other caches of documents that reflect the broad experience of individuals in technology companies and DoD.[3] This would also be useful to ensure that documents are completely equivalent, something that we were unable to do in the present study. Traditional, nontext-based methods that could also be used to examine bottom-up sources include surveys and interviews. Such findings would complement those reported here and provide further verification of the nature of DoD and Silicon Valley cultures. Some additional limitations of the corpus should also be acknowledged. First, the ten-year time frame we used to locate documents could have affected our findings. For example, DoD's recent focus on innovation highlights the way that those in the department speak about their culture changes over time. It is possible that we might have obtained different results if a larger time frame had been exam-

---

[2] C. Pope, "CSAF Outlines Strategic Approach for Air Force Success," U.S. Air Force, August 31, 2020.

[3] V. D. Swain, K. Saha, M. D. Reddy, H. Rajvanshy, G. D. Abowd, and M. De Choudhury, "Modeling Organizational Culture with Workplace Experiences Shared on Glassdoor," Honolulu, Hawaii: *Proceedings of the 2020 CHI Conference on Human Factors in Computing Systems*, April 25–30, 2020; and S. B. Srivastava, A. Goldberg, V. G. Manian, and C. Potts, "Enculturation Trajectories: Language, Cultural Adaptation, and Individual Outcomes in Organizations," *Management Science*, Vol. 64, No. 3, 2018.

ined. Additionally, because some of the articles in the corpus were not written by members of the organizations examined, it is possible that their cultural perceptions differ from organizational members' perceptions. Ultimately, the criterion we used for our corpus represents just one of the criteria that could have been applied. It would be helpful for future studies to employ different criteria to expand on the results reported in this study.

Furthermore, we are not able to fully account for the impact of specific subcultures on the basis of the current project. For instance, research has noted that subcultures often exist in organizations even if the organization can still be characterized by a dominant culture. Within the U.S. military, it is well documented that the various branches of service have unique cultures and that specific units within each branch may also differ from the overall culture of their host organization.[4] To fully understand the culture of the military would require a large-scale assessment of all aspects of DoD (i.e., including the various agencies and branches of each service and their respective subunits) and a comparison of its civilian and noncivilian workforce. Additionally, there is evidence that technology firms (e.g., Google, Amazon, Facebook) differ on several culture dimensions.[5] Although the existence of subcultures cannot be completely accounted for within the current study, it is worth reiterating that organizations can be characterized as possessing a dominant culture.[6] Furthermore, it is unlikely that job applicants make nuanced distinctions about an organization's culture when they evaluate it. For instance, within recruitment contexts in particular, job applicants often rely on cultural stereotypes rather than perceptions of the true organizational culture.[7] Thus, assessing culture from a broader perspective is useful because it is these broad perceptions that applicants (e.g., potential AI workers from the private sector) use to evaluate organizations and inform their impressions. Future research should nonetheless attempt to assess both DoD and Silicon Valley subcultures and determine the extent to which subcultures, as opposed to an organization's dominant culture, affect DoD's ability to engage with Silicon Valley. Knowledge of DoD subcultures could be especially useful and provide insights into who the optimal ambassadors for interacting with this community may be.

An additional limitation of this project concerns the keyword dictionaries that were developed (see Table 4.1). Although these dictionaries were adapted from extant dictionaries[8] and developed from key articles discussing the theoretical foundation of the CVF,[9] it is possible that other words could have been included, which might yield a slightly different pattern

---

[4] Jackson et al., 2018; S. R. Zimmerman, K. Jackson, N. Lander, C. Roberts, D. Madden, and R. Orrie, *Movement and Maneuver: Culture and the Competition for Influence Among the U.S. Military Services*, Santa Monica, Calif.: RAND Corporation, RR-2270-OSD, 2019.

[5] Sull, Sull, and Chamberlain, 2019.

[6] Ostroff, Kinicki, and Muhammad, 2013.

[7] De Goede, Van Vianen, and Klehe, 2011.

[8] Erhardt, 2018.

[9] Hartnell, Ou, and Kinicki, 2011.

of results. Although we believe that these word dictionaries are sufficiently representative of the organizational culture dimensions that they are meant to reflect, it could be beneficial for future research to continue developing and validating these word lists. Having a standardized dictionary for assessing the organizational culture of DoD and Silicon Valley could open the door for many additional text-analytics studies. It would also offer a systematic way to consistently assess organizational culture and ensure that the results are comparable over time. Also, regarding a specific component of the dictionaries, it is possible that the results of the Sense of Duty dimension were slightly biased to favor the DoD corpus because this culture dimension was designed specifically for the military.[10] Although this is a possibility, this simply reinforces the fact that the organizational cultures of DoD and Silicon Valley are unique, and that to understand the cultures of these two organizations, unique dimensions are required. Relatedly, although this project relied on the CVF as a theoretical framework, there are inherent limitations to using these sorts of typologies. For example, the strength of cultural norms and artifacts are not adequately captured within this model via the text analytics approach that we employed. As noted in the introduction, however, any theoretical framework that is employed will necessitate certain trade-offs. An interesting avenue for future research would be to take a bottom-up, rather than top-down, approach to assessing culture. That is, researchers could explore cultural themes that emerge within a corpus, rather than delineating a priori which themes they are examining. Future projects could also potentially employ other theoretical frameworks and/or organizational culture concepts (e.g., tightness-looseness, strength, consensus).

As explained previously, the Sense of Duty dimension was originally designed to capture a unique aspect of DoD. With the rising interest in corporate social responsibility[11] and its growing importance to the recruitment of younger generations of employees,[12] it may be time for researchers to consider creating an equivalent to the Sense of Duty dimension of organizational culture to measure how these initiatives have affected the culture of private sector corporations. Understanding this aspect of corporate culture could become especially important to the technology sector as potential employees of these corporations increasingly question whether they make a positive impact on society.[13] Although any Sense of Duty for a private sector corporation will most likely be different from the concept as applied to DoD, there could be areas of overlap and commonality.

Lastly, we note one final area for future research that would be beneficial for better understanding the organizational culture of DoD and Silicon Valley and would complement the findings reported here. Using the collocate analysis, DoD talks about changing its culture

---

[10] Tinoco and Arnaud, 2013.

[11] M. Gavin, "5 Examples of Corporate Social Responsibility That Were Successful," *Harvard Business School Online*, blog post, June 6, 2019.

[12] Aziz, 2020.

[13] E. Goldberg, "'Techlash' Hits College Campuses," *New York Times*, January 11, 2020.

very frequently and far more often than Silicon Valley. It is unclear, however, whether the culture of DoD is actually changing. Therefore, examining changes in culture over time would be crucial to better understanding whether DoD leadership is succeeding in altering DoD's culture as intended. Measuring this could be accomplished with the use of longitudinal methods (e.g., latent growth curve modeling, cross-lagged panel analysis, multilevel modeling) that employ text analytic approaches, surveys, or some combination thereof.

In conclusion, although there are differences between the organizational cultures of DoD and Silicon Valley, there are also many ways in which these cultures overlap. The text-analytic results reported here provide some preliminary insights into these points of convergence and divergence and offer a valuable lens through which to better understand the civil-military divide in AI. The ability of DoD to recruit the top AI talent from industry and make technological advances is viewed as an increasingly crucial goal of DoD—although this effort could be challenging to accomplish in practice. This project demonstrates that understanding the differences between the organizational cultures of DoD and Silicon Valley is an important consideration for addressing this topic and improving the communication between these two unique communities.

Keyness Analysis for Different Dimensions of Organizational Culture

The tables in this appendix list each of the keywords associated with the five dimensions of organizational culture. They also list how frequently those words appeared in both the DoD corpus of documents and the Silicon Valley corpus of documents, whether they appeared more frequently in the DoD corpus (overpresent) or Silicon Valley corpus (underpresent), and the loglikelihood (*LL*) as a measure how significant the over/under-present result is (higher *LL*s are more impactful).

TABLE A.1

Keyness Analysis: Full Results for Hierarchy Dimension

Word	*LL*	Over/Under	DoD Word Frequency	SV Word Frequency
accountability	16.50	Over	73	35
accountable	53.74	Over	78	14
authorities	21.99	Over	28	4
authority	84.09	Over	91	9
bureaucracy	4.08	Over	29	17
bureaucratic	12.29	Over	27	8
capabilities	76.98	Over	186	60
capability	69.29	Over	117	26
capable	16.69	Over	59	25
communicate	4.09	Under	22	40
communicates	3.10	Over	10	4
communicating	0.06	Under	11	13
communications	2.81	Under	39	59
communication	1.18	Under	51	67
communicators	18.96	Over	13	0

Table A.1—Continued

Word	LL	Over/Under	DoD Word Frequency	SV Word Frequency
consistent	3.01	Over	41	29
consistently	0.03	Under	24	27
control	0.00	Over	119	127
controlled	0.47	Under	9	13
controlling	1.04	Over	13	9
coordinate	1.87	Over	11	6
coordination	7.47	Over	13	3
efficiencies	9.61	Over	15	3
efficiency	1.63	Under	28	41
efficient	7.31	Under	23	48
efficiently	0.99	Over	14	10
formal	4.17	Over	39	25
formally	0.16	Over	8	7
hierarchical	8.06	Over	19	6
hierarchy	9.72	Over	35	15
monitor	0.37	Under	11	15
monitored	0.19	Over	5	4
organization	4.83	Over	254	223
organizational	100.92	Over	188	47
organizations	12.27	Over	184	133
organization's	3.74	Over	22	12
policies	0.03	Over	70	73
policy	13.71	Over	153	103
predict	1.52	Under	9	16
predictability	4.32	Over	6	1
predictable	1.32	Over	10	6
procedures	25.66	Over	46	11
rank	26.57	Over	63	20
ranks	43.89	Over	109	36
regulation	3.02	Over	22	13

Table A.1—Continued

Word	LL	Over/Under	DoD Word Frequency	SV Word Frequency
regulations	41.85	Over	41	3
reliable	2.77	Over	19	11
rule	0.00	Under	26	28
rules	4.23	Under	62	93
stability	20.09	Over	44	13
stable	0.00	Under	12	13
standard	12.27	Over	73	40
standards	73.29	Over	214	80
structure	2.15	Under	67	91
structured	0.61	Under	11	16

NOTES: *Over* refers to words that were conspicuously overpresent (i.e., appeared more) in the target data set (DoD documents) compared with the baseline data set (Silicon Valley documents). *Under* refers to words that were conspicuously underpresent (i.e., appeared less) in the target data set (DoD documents) compared with the baseline data set (Silicon Valley documents). SV = Silicon Valley.

TABLE A.2

Keyness Analysis: Full Results for Adhocracy Dimension

Word	LL	Over/Under	DoD Word Frequency	SV Word Frequency
adaptable	40.85	Over	28	0
adapt	20.57	Over	64	25
adaptability	37.93	Over	26	0
adapted	3.91	Over	11	4
adapting	0.76	Over	10	7
adaptation	10.72	Over	16	3
adaptive	36.78	Over	34	2
agile	11.39	Over	57	29
agility	43.84	Over	67	13
anticipate	3.10	Over	17	9
autonomous	5.51	Over	22	10
autonomy	0.02	Under	15	17
change	27.19	Over	480	359
changed	3.45	Over	61	89

Table A.2—Continued

Word	LL	Over/Under	DoD Word Frequency	SV Word Frequency
changes	5.44	Over	154	125
changing	4.17	Over	101	80
create	30.79	Under	145	272
created	30.57	Under	65	153
creates	0.49	Under	41	51
creating	6.63	Under	95	143
creative	0.00	Under	62	67
creatively	6.62	Over	14	4
creativity	0.00	Under	49	53
detail	7.12	Under	7	22
detailed	3.71	Over	32	20
details	7.36	Under	27	54
dynamic	0.05	Under	29	33
dynamics	19.05	Under	9	40
entrepreneurial	16.08	Under	5	28
entrepreneurs	95.48	Under	6	101
entrepreneurship	16.76	Under	6	31
experiment	1.88	Under	18	29
experimental	0.02	Over	6	6
experimentation	1.97	Over	15	9
experimenting	1.90	Under	5	11
flexibility	0.86	Over	32	27
flexible	1.81	Over	23	16
free	68.85	Under	46	172
freedom	6.44	Under	69	109
freedoms	3.31	Under	7	2
future	102.94	Over	477	234
futures	4.78	Over	12	4
grow	61.15	Under	42	155
growing	43.66	Under	56	158

Table A.2—Continued

Word	LL	Over/Under	DoD Word Frequency	SV Word Frequency
grown	8.12	Under	24	51
growth	288.71	Under	31	348
imagine	17.28	Under	19	57
imagined	0.41	Under	6	9
innovate	14.37	Under	25	63
innovating	1.27	Under	14	22
innovation	0.08	Under	296	325
innovations	17.37	Under	14	48
innovative	11.13	Over	127	86
innovators	5.03	Over	32	18
new	189.65	Under	685	1376
opportunities	0.01	Over	132	140
opportunity	5.93	Under	138	194
research	2.96	Under	92	125
risk	66.54	Over	187	68
risks	1.14	Over	51	44
risky	0.21	Under	5	7
stimulate	0.19	Over	5	4
temporarily	3.10	Over	10	4
uncertain	20.81	Over	27	4
uncertainties	1.55	Over	5	2
uncertainty	12.46	Over	37	14
variety	0.51	Under	28	36
vision	31.58	Under	48	127

NOTES: *Over* refers to words that were conspicuously overpresent (i.e., appeared more) in the target data set (DoD documents) compared with the baseline data set (Silicon Valley documents). *Under* refers to words that were conspicuously underpresent (i.e., appeared less) in the target data set (DoD documents) compared with the baseline data set (Silicon Valley documents). SV = Silicon Valley.

TABLE A.3

Keyness Analysis: Full Results for Clan Dimension

Word	LL	Over/Under	DoD Word Frequency	SV Word Frequency
attached	0.84	Over	9	6
cohesion	19.68	Over	21	2
cohesive	0.93	Over	8	5
cohesiveness	7.29	Over	5	0
collaborate	0.05	Under	12	14
collaboration	30.91	Under	34	102
collaborative	1.99	Under	17	28
competence	105.46	Over	95	5
competencies	15.72	Over	15	1
competency	4.78	Over	12	4
competent	4.29	Over	16	7
concern	13.96	Over	45	18
concerned	2.05	Over	35	26
concerning	12.19	Over	15	2
concerns	2.85	Under	35	54
consensus	0.63	Under	7	11
developing	4.61	Over	97	75
develop	42.26	Over	208	105
developed	1.90	Over	75	79
development	42.30	Over	289	167
developmental	60.21	Over	58	4
develops	17.14	Over	19	2
empower	11.89	Under	20	51
empowered	0.01	Over	20	21
empowering	4.99	Under	20	24
empowerment	3.53	Under	9	20
empowers	0.18	Over	6	5
human resources	99.14	Under	8	111
individual	54.75	Over	195	83

Table A.3—Continued

Word	LL	Over/Under	DoD Word Frequency	SV Word Frequency
individually	13.42	Over	16	2
individuals	15.69	Over	97	54
involve	6.59	Over	22	9
involved	8.70	Over	69	42
involves	0.06	Over	23	23
involving	0.71	Over	11	8
involvement	8.06	Over	19	6
loyal	0.93	Over	8	5
loyalty	53.57	Over	98	24
mentor	17.71	Over	36	10
mentoring	15.03	Over	22	4
mentors	9.61	Over	15	3
mentorship	37.92	Over	38	3
morale	7.37	Over	26	11
participation	2.28	Over	13	7
relationship	6.68	Over	79	54
relationships	5.56	Over	67	46
satisfaction	0.05	Over	18	18
self-development	7.29	Over	5	0
skill	9.13	Over	44	22
skills	5.46	Over	153	124
support	101.62	Over	375	163
supported	1.56	Over	28	21
supporting	15.84	Over	41	14
supportive	3.49	Over	12	5
supports	2.26	Under	15	26
team	111.18	Under	245	570
teammates	0.16	Under	6	8
teams	14.08	Under	127	207

Table A.3—Continued

Word	LL	Over/Under	DoD Word Frequency	SV Word Frequency
teamwork	13.33	Under	13	41
train	48.05	Over	75	15
trained	72.74	Over	82	9
training	631.80	Over	648	55
trains	1.66	Over	8	4
trust	76.77	Over	286	125
trusted	5.66	Over	42	25
voice	28.33	Under	27	86

NOTES: *Over* refers to words that were conspicuously overpresent (i.e., appeared more) in the target data set (DoD documents) compared with the baseline data set (Silicon Valley documents). *Under* refers to words that were conspicuously underpresent (i.e., appeared less) in the target data set (DoD documents) compared with the baseline data set (Silicon Valley documents). SV = Silicon Valley.

TABLE A.4

Keyness Analysis: Full Results for Market Dimension

Word	LL	Over/Under	DoD Word Frequency	SV Word Frequency
achievement	3.10	Over	17	9
achievable	3.20	Over	5	1
achieve	1.85	Over	101	89
achieved	0.16	Over	28	27
achievements	1.94	Over	7	3
achieving	25.10	Over	58	18
aggressively	1.48	Over	21	15
aggression	27.72	Over	19	0
aggressive	4.37	Under	23	42
communicate	4.09	Under	22	40
communicates	3.10	Over	10	4
communicating	0.06	Under	11	13
communications	2.81	Under	39	59
communications	1.88	Under	51	67
communicators	18.96	Over	13	0
competence	105.46	Over	95	5

Table A.4—Continued

Word	LL	Over/Under	DoD Word Frequency	SV Word Frequency
competencies	15.72	Over	15	1
competency	4.78	Over	12	4
competent	4.29	Over	16	7
competing	5.15	Under	17	35
competition	12.30	Under	61	113
competitions	3.20	Over	5	1
competitiveness	0.03	Under	5	6
competitive	5.34	Under	60	94
competitors	70.44	Under	11	95
contract	3.55	Over	40	27
contracting	11.77	Over	12	1
contractor	1.55	Over	5	27
contractors	7.75	Over	25	10
contracts	0.53	Over	18	15
control	0.00	Over	119	127
controlled	0.47	Under	9	13
controlling	1.04	Over	13	9
coordinate	1.87	Over	11	6
customer	378.80	Under	13	690
customers	832.57	Under	8	359
energy	1.51	Over	91	81
environment	32.91	Over	230	134
environments	30.93	Over	53	12
external	0.03	Under	23	29
fast	87.14	Under	21	137
faster	34.43	Under	23	86
goal	10.56	Over	143	101
goals	22.28	Over	151	87
market	316.22	Over	17	324
markets	50.05	Over	11	76

Table A.4—Continued

Word	LL	Over/Under	DoD Word Frequency	SV Word Frequency
perform	9.35	Over	55	30
performance	36.93	Over	363	236
performed	0.03	Over	12	12
performers	14.64	Under	10	37
performing	7.77	Over	28	12
plan	4.55	Over	125	101
planned	0.01	Under	18	20
planners	10.48	Over	11	1
planning	3.91	Over	75	57
plans	4.45	Over	91	70
productive	18.26	Under	8	37
productivity	48.15	Under	19	93
profit	42.36	Under	3	46
profitable	28.37	Under	1	27
profits	39.49	Under	2	40
rapid	2.31	Under	7	42
rapidly	1.69	Under	31	45
result	0.08	Over	96	99
results	1.08	Under	175	209
return	1.91	Under	40	57
speed	2.24	Under	35	52

NOTES: *Over* refers to words that were conspicuously overpresent (i.e., appeared more) in the target data set (DoD documents) compared with the baseline data set (Silicon Valley documents). *Under* refers to words that were conspicuously underpresent (i.e., appeared less) in the target data set (DoD documents) compared with the baseline data set (Silicon Valley documents). SV = Silicon Valley.

TABLE A.5

Keyness Analysis: Full Results for Sense of Duty Dimension

Word	LL	Over/Under	DoD Word Frequency	SV Word Frequency
allegiance	1.94	Over	7	3
authority	84.09	Over	91	9
cause	9.31	Over	47	24
conduct	100.32	Over	158	32
courage	94.24	Over	122	18
courageous	2.07	Over	10	5
courageously	8.75	Over	6	0
duty	206.01	Over	159	3
duties	25.63	Over	40	8
devotion	3.31	Over	7	2
discipline	23.91	Over	55	17
disciplined	18.9	Over	41	12
disciplinary	5.5	Over	7	1
disciplines	3.2	Over	5	1
honor	456.19	Over	367	10
honorable	65.64	Over	45	0
honored	9.2	Over	10	1
honoring	5.5	Over	7	1
honors	0.01	Over	5	5
integrity	55.54	Over	98	23
loyalty	53.57	Over	98	24
loyal	0.93	Over	8	5
mission	283.36	Over	606	175
obey	10.21	Over	7	0
purpose	8.87	Over	118	83
purposes	3.73	Over	18	9
respect	89.48	Over	182	50
respectful	9.61	Over	17	4
respectively	8.75	Over	6	0

Table A.5—Continued

Word	LL	Over/Under	DoD Word Frequency	SV Word Frequency
respective	6.87	Over	16	5
respects	2.39	Over	6	2
respected	0.64	Over	13	10
sacrifice	33.77	Over	45	7
sacrifices	20.96	Over	22	2
sacrificed	6.71	Over	8	1
sacrificing	0.49	Over	7	5
selfless	23.54	Over	24	2
selflessly	12.21	Over	7	0
service	233.06	Over	743	295
subordinate	173.09	Over	148	6
subordinates	87.08	Over	78	4

NOTES: *Over* refers to words that were conspicuously overpresent (i.e., appeared more) in the target data set (DoD documents) compared with the baseline data set (Silicon Valley documents). *Under* refers to words that were conspicuously underpresent (i.e., appeared less) in the target data set (DoD documents) compared with the baseline data set (Silicon Valley documents). SV = Silicon Valley.

Nonsignificant Stance Comparison Analysis Categories

Table B.1 lists the categories that were found to be nonsignificant when conducting stance comparison analysis (as described in the "Stance Comparison Analysis" section in Chapter Four). See Figure 4.3 in Chapter Four for the list of categories that were found to be significant for this analysis.

TABLE B.1

Nonsignificant Stance Comparison Analysis Categories

List of Nonsignificant Language Categories for SCA		
Abstract concepts	Error recovery	Positive attribution
Agreement	Example	Procedures
Anger	Exceptions	Prohibition
Apology	Fear	Promises
Attention grab	Follow up	Reinforce
Autobiography	Future question	Reluctance
Causality	Generalization	Request
Citing precedent	Imperatives	Sequence
Communicator role	Innovation	Speculative sources
Confirming opinions	Linguistic references	Supporting reasoning
Confront	Negative attribution	Time date
Contested source	Personal reluctance	Undermining sources
Countering sources	Personal thinking	You reference
Dialog cues	Popular opinions	Positive attribution

Abbreviations

AI	artificial intelligence
ASA	attraction-selection-attrition
CVF	Competing Values Framework
DARPA	Defense Advanced Research Projects Agency
DoD	U.S. Department of Defense
LL	loglikelihood
LR	likelihood ratio
NSA	National Security Agency
PMI	Pointwise Mutual Information
SCA	stance comparison analysis

References

Amazon Jobs, "Leadership Principles," webpage, undated. As of February 17, 2022: https://www.amazon.jobs/en/principles

Aziz, Afdhel, "The Power of Purpose: The Business Case for Purpose," *Forbes*, March 7, 2020. As of April 8, 2022: https://www.forbes.com/sites/afdhelaziz/2020/03/07/the-power-of-purpose-the-business-case-for-purpose-all-the-data-you-were-looking-for-pt-2/?sh=46c000a93cf7

Barrick, M. R., and L. Parks-Leduc, "Selection for Fit," *Annual Review of Organizational Psychology and Organizational Behavior*, Vol. 6, 2019, pp. 171–193.

Bezos, J. P., "Letter to Shareholders," AboutAmazon.com, 1997. As of February 17, 2022: https://s2.q4cdn.com/299287126/files/doc_financials/annual/Shareholderletter97.pdf

Breaugh, J. A., "Employee Recruitment: Current Knowledge and Important Areas for Future Research," *Human Resource Management Review*, Vol. 18, No. 3, September 2008, pp. 103–118.

Browne, Ryan, "Top US General Says Google Is 'Indirectly Benefiting the Chinese Military,'" CNN, March 14, 2019. As of June 30, 2021: https://www.cnn.com/2019/03/14/politics/dunford-china-google/index.html

Cameron, K. S., and R. E. Quinn, *Diagnosing and Changing Organizational Culture: Based on the Competing Values Framework*, San Francisco, Calif.: Jossey-Bass, 2006.

Cameron, K., *An Introduction to the Competing Values Framework*, white paper, Holland, Mich.: Haworth, 2009.

Carnegie Mellon University, Department of English, "DocuScope: Computer-Aided Rhetorical Analysis" webpage, undated. As of February 22, 2022: https://www.cmu.edu/dietrich/english/research-and-publications/docuscope.html

Chatman, J. A., and C. A. O'Reilly, "Paradigm Lost: Reinvigorating the Study of Organizational Culture," *Research in Organizational Behavior*, Vol. 36, 2016, pp. 199–224.

Cohen, J., *Statistical Power Analysis for the Behavioral Sciences*, Abingdon, England: Routledge Academic, 1988.

Cole, R. F., "Understanding Military Culture: A Guide for Professional School Counselors," *Professional Counselor*, Vol. 4, No. 5, 2014, pp. 497–504.

Coll, J. E., E. L. Weiss, P. Draves, and D. Dyer, "The Impact of Military Cultural Awareness, Experience, Attitudes, and Education on Clinician Self-Efficacy in the Treatment of Veterans," *Journal of International Continuing Social Work Education*, Vol. 15, No. 1, January 2012, pp. 39–48.

Cooper, L., N. Caddick, L. Godier, A. Cooper, and M. Fossey, "Transition from the Military into Civilian Life: An Exploration of Cultural Competence," *Armed Forces & Society*, Vol. 44, No. 1, January 2018, pp. 156–177.

Coser, L. A., *Greedy Institutions: Patterns of Undivided Commitment*, New York: Free Press, 1974.

Cox, A. B., "Mechanisms of Organizational Commitment: Adding Frames to Greedy Institution Theory," *Sociological Forum*, Vol. 31, No. 3, May 2016, pp. 685–708.

De Goede, M. E., A. E. Van Vianen, and U. C. Klehe, "Attracting Applicants on the Web: PO Fit, Industry Culture Stereotypes, and Website Design," *International Journal of Selection and Assessment*, Vol. 19, No. 1., February 2011, pp. 51–61.

Denison, D. R., and A. K. Mishra, "Toward a Theory of Organizational Culture and Effectiveness," *Organization Science*, Vol. 6, No. 2, April 1995, pp. 204–223.

Department of the Navy, "Department of the Navy Core Values Charter," webpage, undated. As of February 17, 2022:
https://www.secnav.navy.mil/ethics/pages/corevaluescharter.aspx

Dokko, G., and W. Jiang, "Managing Talent Across Organizations: The Portability of Individual Performance," in D. G. Collings, K. Mellahi, and W. F. Cascio, eds., *The Oxford Handbook of Talent Management*, Oxford, UK: Oxford University Press, 2017, pp. 115–133.

Edwards, J. R., and D. M. Cable, "The Value of Value Congruence," *Journal of Applied Psychology*, Vol. 94, No. 3, June 2009, pp. 654–677.

Erhardt, R., *Cultural Analysis of Organizational Development Units: A Comprehensive Approach Based on the Competing Values Framework*, dissertation, Atlanta, Ga.: Georgia State University, 2018.

Everstine, B. W., "Esper: Culture Change in DoD Needed to Improve Acquisition Process," *Air Force Magazine*, January 2020. As of February 17, 2022:
https://www.airforcemag.com/esper-culture-change-in-dod-needed-to-improve-acquisition-process/

Finkle, T. A., "Corporate Entrepreneurship and Innovation in Silicon Valley: The Case of Google, Inc.," *Entrepreneurship Theory and Practice*, Vol. 36, No. 4, July 2012, pp. 863–887.

Fritz, C. O., P. E. Morris, and J. J. Richler, "Effect Size Estimates: Current Use, Calculations, and Interpretation," *Journal of Experimental Psychology: General*, Vol. 141, No. 1, August 2011, pp. 2–18.

Gardner, W. L., B. J. Reithel, C. C. Cogliser, F. O. Walumbwa, and R. T. Foley, "Matching Personality and Organizational Culture: Effects of Recruitment Strategy and the Five-Factor Model on Subjective Person–Organization Fit," *Management Communication Quarterly*, Vol. 26, No. 4, July 2012, pp. 585–622.

Gavin, M., "5 Examples of Corporate Social Responsibility That Were Successful," *Harvard Business School Online*, blog post, June 6, 2019. As of February 22, 2022:
https://online.hbs.edu/blog/post/corporate-social-responsibility-examples

Gerras, S. J., L. Wong, and C. D. Allen, *Organizational Culture: Applying a Hybrid Model to the U.S. Army*, Carlisle Barracks, Pa.: U.S. Army War College, 2008.

Goldberg, E., "'Techlash' Hits College Campuses," *New York Times*, January 11, 2020.

Goodwin, G.F., N. Blacksmith, and M. R. Coats, "The Science of Teams in the Military: Contributions From Over 60 Years of Research," *American Psychologist*, Vol. 73, No. 4, 2018, pp. 322–333.

Google, "Ten Things We Know To Be True," webpage, undated. As of February 17, 2022:
https://www.google.com/about/philosophy.html

Gordon, G. G., "Industry Determinants of Organizational Culture," *Academy of Management Review*, Vol. 16, No. 2, April 1991, pp. 396–415.

Greene, T., J. Buckman, C. Dandeker, and N. Greenberg, "The Impact of Culture Clash on Deployed Troops," *Military Medicine*, Vol. 175, No. 12, December 2010, pp. 958–963.

Gregory, B. T., S. G. Harris, A. A. Armenakis, and C. L. Shook, "Organizational Culture and Effectiveness: A Study of Values, Attitudes, and Organizational Outcomes," *Journal of Business Research*, Vol. 62, No. 7, July 2009, pp. 673–679.

Grinstein, A., and A. Goldman, "Characterizing the Technology Firm: An Exploratory Study," *Research Policy*, Vol. 35, No. 1, February 2006, pp. 121–143.

Groysberg, B., L. E. Lee, and A. Nanda, "Can They Take It with Them? The Portability of Star Knowledge Workers' Performance," *Management Science*, Vol. 54, No. 7, July 2008, pp. 1213–1230.

Groysberg, B., J. Lee, J. Price, and J. Cheng, "The Leader's Guide to Corporate Culture," *Harvard Business Review*, Vol. 96, January–February 2018, pp. 44–52.

Guiso, L., P. Sapienza, and L. Zingales, "The Value of Corporate Culture," *Journal of Financial Economics*, Vol. 117, No. 1, July 2015, pp. 60–76.

Gupta, B., "A Comparative Study of Organizational Strategy and Culture Across Industry," *Benchmarking: An International Journal*, Vol. 18, No. 4, July 2011, pp. 510–528.

Hartnell, C. A., A. Y. Ou, and A. Kinicki, "Organizational Culture and Organizational Effectiveness: A Meta-Analytic Investigation of the Competing Values Framework's Theoretical Suppositions," *Journal of Applied Psychology*, Vol. 96, No. 4, July 2011, pp. 677–694.

Hartnell, C. A., A. Y. Ou, A. Kinicki, D. Choi, and E. P. Karam, "A Meta-Analytic Test of Organizational Culture's Association with Elements of an Organization's System and its Relative Predictive Validity on Organizational Outcomes," *Journal of Applied Psychology*, Vol. 104, No. 6, June 2019, pp. 832–850.

Irving, D., "Big Data, Big Questions," *RAND Review*, October 16, 2017. As of August 1, 2021: https://www.rand.org/blog/rand-review/2017/10/big-data-big-questions.html

Jackson, K., K. L. Kidder, S. Mann, W. H. Waggy II, N. Lander, and S. R. Zimmerman, *Raising the Flag: Implications of U.S. Military Approaches to General and Flag Officer Development*, Santa Monica, Calif.: RAND Corporation, RR-4347-OSD, 2020. As of February 16, 2022: https://www.rand.org/pubs/research_reports/RR4347.html

Judge, T. A., and D. M. Cable, "Applicant Personality, Organizational Culture, and Organization Attraction," *Personnel Psychology*, Vol. 50, No. 2, December 2006, pp. 359–394.

Jung, T., T. Scott, H. T. Davies, P. Bower, D. Whalley, R. McNally, and R. Mannion, "Instruments for Exploring Organizational Culture: A Review of the Literature," *Public Administration Review*, Vol. 69, No. 6, November/December 2009, pp. 1087–1096.

Kania, E. B., and E. Moore, "Great Power Rivalry Is Also a War for Talent," *Defense One*, May 19, 2019. As of February 17, 2022: https://www.defenseone.com/ideas/2019/05/great-power-rivalry-also-war-talent/157103/

Kavanagh, J., W. Marcellino, J. S. Blake, S. Smith, S. Davenport, and M. Gizaw, *News in a Digital Age: Comparing the Presentation of News Information over Time and Across Media Platforms*, Santa Monica, Calif.: RAND Corporation, RR-2960-RC, 2019. As of February 20, 2022: https://www.rand.org/pubs/research_reports/RR2960.html

Kepe, M., "Considering Military Culture and Values When Adopting AI," *Small Wars Journal*, June 15, 2020. As of February 17, 2022: https://smallwarsjournal.com/jrnl/art/considering-military-culture-values-when-adopting-ai

Khosla, V., "The Silicon Valley Culture," *Medium*, January 17, 2018. As of February 22, 2022: https://medium.com/@vkhosla/the-silicon-valley-culture-bdc86db0b524

Kobayashi, V. B., S. T. Mol, H. A. Berkers, G. Kismihók, and D. N. Den Hartog, "Text Mining in Organizational Research," *Organizational Research Methods*, Vol. 21, No. 3, July 2018, pp. 733–765.

Kristof-Brown, A. L., R. D. Zimmerman, and E. C. Johnson, "Consequences of Individuals' Fit at Work: A Meta-Analysis of Person-Job, Person-Organization, Person-Group, and Person-Supervisor Fit," *Personnel Psychology*, Vol. 58, No. 2, 2005, pp. 281–342.

Mansoor, P. R., and W. Murray, eds., *The Culture of Military Organizations*, New York: Cambridge University Press, 2019.

Mattox, J. M., "Values Statements and the Profession of Arms: A Reevaluation," *Joint Force Quarterly*, Issue 68, First Quarter 2013, pp. 59–63.

Marcellino, W. M., K. Cragin, J. Mendelsohn, A. M. Cady, M. Magnuson, and K. Reedy, "Measuring the Popular Resonance of Daesh's Propaganda," *Journal of Strategic Security*, Vol. 10, 2017, pp. 32–52.

Martin, J., *Organizational Culture: Mapping the Terrain*, Thousand Oaks, Calif.: SAGE Publications, 2002.

Martins, H., Y. B. Dias, and S. Khanna, "What Makes Silicon Valley Companies So Successful," *Harvard Business Review*, April 26, 2016. As of February 22, 2022: https://hbr.org/2016/04/what-makes-some-silicon-valley-companies-so-successful

Mattis, J., *Summary of the 2018 National Defense Strategy of the United States of America*, Washington, D.C.: U.S. Department of Defense, 2018.

Mawdsley, J. K., and D. Somaya, "Employee Mobility and Organizational Outcomes: An Integrative Conceptual Framework and Research Agenda," *Journal of Management*, Vol. 42, No. 1, 2016, pp. 85–113.

McCormick, W., J. Currier, S. Isaak, B. Sims, B. Slagel, T. Carroll, K. Hammer, and D. Albright, "Military Culture and Post-Military Transitioning Among Veterans: A Qualitative Analysis," *Journal of Veterans Studies*, Vol. 4, No. 2, August 2019, pp. 287–298.

Mehta, A., "Cultural Divide: Can the Pentagon Crack Silicon Valley?" *Defense News*, January 28, 2019. As of February 17, 2022: https://www.defensenews.com/pentagon/2019/01/28/cultural-divide-can-the-pentagon-crack-silicon-valley/

Meredith, L. S., C. S. Sims, B. S. Batorsky, A. T. Okunogbe, B. L. Bannon, and C. A. Myatt, *Identifying Promising Approaches to U.S. Army Institutional Change: A Review of the Literature on Organizational Culture and Climate*, Santa Monica, Calif.: RAND Corporation, RR-1588-A, 2017. As of February 17, 2022: https://www.rand.org/pubs/research_reports/RR1588.html

Meta Careers, "Culture at Meta," webpage, undated. As of February 17, 2022: https://www.metacareers.com/facebook-life/

Metz, Cade, "Pentagon Wants Silicon Valley's Help on AI," *New York Times*, March 15, 2018.

Meyer, E. G., J. E. McCarroll, and R. J. Ursano, eds., *U.S. Army Culture: An Introduction for Behavioral Health Researchers*, Bethesda, Md.: Center for the Study of Traumatic Stress, 2017.

Microsoft, "Company Values," webpage, undated-a. As of February 17, 2022: https://www.microsoft.com/en-us/about/values

Microsoft, "Corporate Values," webpage, undated-b. As of February 17, 2022: https://www.microsoft.com/en-us/about/corporate-values

Milburn, A., "Losing Small Wars: Why US Military Culture Leads to Defeat," *Small Wars Journal*, September 12, 2021. As of February 17, 2022:
https://smallwarsjournal.com/jrnl/art/losing-small-wars-why-us-military-culture-leads-defeat

Military Leadership Diversity Commission, "Department of Defense Core Values," Washington, D.C., Issue Paper No. 6, December 2009.

Miller, J., "In Goodbye Message, Chaillan Unloads His Frustrations over DoD's Technology Culture, Processes," *Federal News Network*, September 2, 2021. As of February 17, 2022:
https://federalnewsnetwork.com/people/2021/09/in-goodbye-message-chaillan-unloads-his-frustrations-over-dods-technology-culture-processes/

Müller, S. D., and P. A. Nielsen, "Competing Values in Software Process Improvement: A Study of Cultural Profiles," *Information Technology & People*, Vol. 26, November 2013, pp. 146–171.

Murphy, M. G., and K. M. Davey, "Ambiguity, Ambivalence and Indifference in Organisational Values," *Human Resource Management Journal*, Vol. 12, No. 1, 2002, pp. 17–32.

National Security Commission on Artificial Intelligence, *Final Report of the National Security Commission on Artificial Intelligence*, Arlington, Va., March 2021.

Naval History and Heritage Command, "Sailor's Creed," webpage, October 16, 2018. As of February 17, 2022:
https://www.history.navy.mil/browse-by-topic/heritage/customs-and-traditions0/the-sailor-s-creed.html

Netflix, "Netflix Culture," webpage, undated. As of February 17, 2022:
https://jobs.netflix.com/culture

Office of the Director of Administration and Management, U.S. Department of Defense, "Organizational Assessment," webpage, undated. As of March 7, 2022:
https://dam.defense.gov/Publications/Organizational-Assessment/

O'Reilly, C. A., III, J. Chatman, and D. F. Caldwell, "People and Organizational Culture: A Profile Comparison Approach to Assessing Person-Organization Fit," *Academy of Management Journal*, Vol. 34, No. 3, 1991, pp. 487–516.

Ostroff, C., A. J. Kinicki, and R. S. Muhammad, "Organizational Culture and Climate," in I. B. Weiner, ed., *Handbook of Psychology*, 2nd ed., Hoboken, N.J.: John Wiley & Sons, 2013, pp. 643–676.

Pandey, S., and S. K. Pandey, "Applying Natural Language Processing Capabilities in Computerized Textual Analysis to Measure Organizational Culture," *Organizational Research Methods*, Vol. 22, No. 3, 2019, pp. 765–797.

Pardes, A., "Silicon Valley Ruined Work Culture," *Wired*, February 24, 2020. As of February 17, 2022:
https://www.wired.com/story/how-silicon-valley-ruined-work-culture/

Pasch, S., *Corporate Culture and Industry-Fit: A Text Mining Approach*, Institute of Labour Economics, conference paper, Bonn, Germany, September 21–22, 2018.

Pierce, J. G., *Is the Organizational Culture of the U.S. Army Congruent with the Professional Development of its Senior Level Officer Corps?* Carlisle Barracks, Pa.: U.S. Army War College, 2010.

Pinker, S., *The Stuff of Thought: Language as a Window into Human Nature*, New York: Viking, 2007.

Ployhart, R. E., D. Hale Jr., and M. C. Campion, "Staffing Within the Social Context," in B. Schneider and K. M. Barbara, eds., *The Oxford Handbook of Organizational Climate and Culture: Antecedents, Consequences, and Practice*, Oxford, UK: Oxford University Press, 2014, pp. 23–43.

Pollman, A., *Diagnosis and Analysis of Marine Corps Organizational Cultural*, Executive Master of Business Administration Capstone Project Report, Monterey, Calif.: Naval Postgraduate School, March 2015.

Pollman, A., "Framing Marine Corps Culture," *Proceedings*, U.S. Naval Institute, June 2018. As of February 17, 2022:
https://www.usni.org/magazines/proceedings/2018/june/framing-marine-corps-culture

Pope, C., "CSAF Outlines Strategic Approach for Air Force Success," U.S. Air Force, August 31, 2020.

Quinn, R. E., and J. Rohrbaugh, "A Spatial Model of Effectiveness Criteria: Towards a Competing Values Approach to Organizational Analysis," *Management Science*, Vol. 29, No. 3, March 1983, pp. 363–377.

Redmond, S. A., S. L. Wilcox, S. Campbell, A. Kim, K. Finney, K. Barr, and A. M. Hassan, "A Brief Introduction to the Military Workplace Culture," *Work*, Vol. 50, No. 1, 2015, pp. 9–20.

Roberge, M. É., and R. Van Dick, "Recognizing the Benefits of Diversity: When and How Does Diversity Increase Group Performance?" *Human Resource Management Review*, Vol. 20, No. 4, 2010, pp. 295–308.

Robers, B., "Public Understanding of the Profession of Arms," *Military Review*, November–December 2012.

Rock, D., and H. Grant, "Why Diverse Teams Are Smarter," *Harvard Business Review*, November 4, 2016.

Ryseff, J., "How to (Actually) Recruit Talent for the AI Challenge," *War on the Rocks*, February 5, 2020. As of February 17, 2022:
https://warontherocks.com/2020/02/how-to-actually-recruit-talent-for-the-ai-challenge/

Sargent, J., and M. Gallo, *The Global Research and Development Landscape and Implications for the Department of Defense*, Washington, D.C.: Congressional Research Service, R45403, June 28, 2021.

Schein, E. H., *Organizational Culture and Leadership*, San Francisco, Calif.: Jossey-Bass, 2010.

Schmiedel, T., O. Müller, and J. vom Brocke, "Topic Modeling as a Strategy of Inquiry in Organizational Research: A Tutorial with an Application Example on Organizational Culture," *Organizational Research Methods*, Vol. 22, 2019, pp. 941–968.

Schneider, B., "The People Make the Place," *Personnel Psychology*, Vol. 40, 1987, pp. 437–453.

Schneider, B., M. G. Ehrhart, and W. H. Macey, "Organizational Climate and Culture," *Annual Review of Psychology*, Vol. 64, 2013, pp. 361–388.

Segal, M. W., "The Military and the Family as Greedy Institutions," *Armed Forces & Society*, Vol. 13, No. 1, 1986, pp. 9–38.

Select Committee on Artificial Intelligence of the National Science and Technology Council, *The National Artificial Intelligence Research and Development Strategic Plan: 2019 Update*, Washington, D.C.: Executive Office of the President, 2019.

Shane, Scott, and Daisuke Wakabayashi, "'The Business of War': Google Employees Protest Work for the Pentagon," *New York Times*, April 4, 2018.

Short, J. C., A. F. McKenny, and S. W. Reid, "More Than Words? Computer-Aided Text Analysis in Organizational Behavior and Psychology Research," *Annual Review of Organizational Psychology & Organizational Behavior*, Vol. 5, No. 1, 2018, pp. 415–435.

Sørlie, H. O., J. Hetland, A. Dysvik, T. H. Fosse, and Ø. L. Martinsen, "Person-Organization Fit in a Military Selection Context," *Military Psychology*, Vol. 32, No. 3, 2020, pp. 237–246.

Srivastava, S. B., and A. Goldberg, "Language as a Window into Culture," *California Management Review*, Vol. 60, No. 1, 2017, pp. 56–69.

Srivastava, S. B., A. Goldberg, V. G. Manian, and C. Potts, "Enculturation Trajectories: Language, Cultural Adaptation, and Individual Outcomes in Organizations," *Management Science*, Vol. 64, No. 3, 2018, pp. 1348–1364.

Sull, D., C. Sull, and A. Chamberlain, *Measuring Culture in Leading Companies*, Cambridge, Mass.: MIT Sloan Management Review, 2019.

Sullivan, S. E., and Y. Baruch, "Advances in Career Theory and Research: A Critical Review and Agenda for Future Exploration," *Journal of Management*, Vol. 35, No. 6, 2009, pp. 1542–1571.

Swain, V. D., K. Saha, M. D. Reddy, H. Rajvanshy, G. D. Abowd, and M. De Choudhury, "Modeling Organizational Culture with Workplace Experiences Shared on Glassdoor," Honolulu, Hawaii: *Proceedings of the 2020 CHI Conference on Human Factors in Computing Systems*, April 25–30, 2020.

Swain, R. M., and A. C. Pierce, *The Armed Forces Officer*, Washington, D.C.: National Defense University Press, 2017.

Szayna, T. S., E V. Larson, A. O'Mahony, S. Robson, A. Gereben Schaefer, M. Matthews, J. M. Polich, L. Ayer, D. Eaton, W. Marcellino, L. Kraus, M. N. Posard, J. Syme, Z. Winkelman, C. Wright, M. Zander Cotugno, and W. Welser IV, *Considerations for Integrating Women into Closed Occupations in U.S. Special Operations Forces*, Santa Monica, Calif.: RAND Corporation, RR-1058-USSOCOM, 2016. As of February 18, 2022:
https://www.rand.org/pubs/research_reports/RR1058.html

Tarraf, D. C., W. Shelton, E. Parker, B. Alkire, D. Gehlhaus, J. Grana, A. Levedahl, J. Léveillé, J. Mondschein, J. Ryseff, A. Wyne, D. Elinoff, E. Geist, B. N. Harris, E. Hui, C. Kenney, S. Newberry, C. Sachs, P. Schirmer, D. Schlang, V. M. Smith, A. Tingstad, P. Vedula, and K. Warren, *The Department of Defense Posture for Artificial Intelligence: Assessment and Recommendations*, Santa Monica, Calif.: RAND Corporation, RR-4229-OSD, 2019. As of February 18, 2022:
https://www.rand.org/pubs/research_reports/RR4229.html

Tausczik, Y. R., and J. W. Pennebaker, "The Psychological Meaning of Words: LIWC and Computerized Text Analysis Methods," *Journal of Language and Social Psychology*, Vol. 29, No. 1, 2010, pp. 24–54.

Tinoco, J. K., and A. Arnaud, "The Transfer of Military Culture to Private Sector Organizations: A Sense of Duty Emerges," *Journal of Organizational Culture, Communications and Conflict*, Vol. 17, No. 2, 2013, pp. 37–62.

U.S. Air Force, "Vision and Creed," webpage, undated. As of February 17, 2022:
https://www.airforce.com/mission/vision

U.S. Army, "The Army Values," webpage, undated-a. As of February 17, 2022:
https://www.army.mil/values/

U.S. Army, "Soldier's Creed," webpage, undated-b. As of February 17, 2022:
https://www.army.mil/values/soldiers.html

U.S. Department of Veterans Affairs, *The Military to Civil Transition: A Review of Historical, Current, and Future Trends*, Washington, D.C., 2018.

Waltzman, R., L. Ablon, C. Curriden, G. S. Hartnett, M. A. Holliday, L. Ma, B. Nichiporuk, A. Scobell, and D. C. Tarraf, *Maintaining the Competitive Advantage in Artificial Intelligence and Machine Learning*, Santa Monica, Calif.: RAND Corporation, RR-A200-1, 2020. As of February 17, 2022:
https://www.rand.org/pubs/research_reports/RRA200-1.html

Williams, T. M., "Practicing What We Preach: Creating a Culture to Support Mission Command," *Small Wars Journal*, blog post, July 2019. As of February 17, 2022:
https://smallwarsjournal.com/jrnl/art/practicing-what-we-preach-creating-culture-support-mission-command

Wilson, P. H., "Defining Military Culture," *Journal of Military History*, Vol. 72, No. 1, 2008, pp. 11–41.

Wood, C. J., "Marine Corps Innovation: The Need for a Reawakening," *Marine Corps Gazette*, Vol. 99, No. 11, 2015, pp. 33–36.

Zimmerman, S. Rebecca, Kimberly Jackson, Natasha Lander, Colin Roberts, Dan Madden, and Rebeca Orrie, *Movement and Maneuver: Culture and the Competition for Influence Among the U.S. Military Services*, Santa Monica, Calif.: RAND Corporation, RR-2270-OSD, 2019. As of February 16, 2022:
https://www.rand.org/pubs/research_reports/RR2270.html